PRINCIPLES OF FUNGAL TAXONOMY

PRINCIPLES OF FUNGAL TAXONOMY

P. H. B. TALBOT, Ph.D. (London)

Reader in Mycology
Waite Agricultural Research Institute
University of Adelaide
South Australia

Macmillan

First published 1971

Published by
THE MACMILLAN PRESS
London and Basingstoke
Associated companies in New York Toronto Dublin Melbourne Johannesburg and Madras

SBN 333 11561 9 (cased)
333 11564 3 (paper)

Printed in Great Britain by
ROBERT MACLEHOSE AND CO LTD
The University Press, Glasgow

FOREWORD

The aim of this book is to give a concise account of fungi, suitable for a short undergraduate course in mycology. The main problem with such a course is to condense the material into a balanced account, limited enough to fit the time yet broad enough to avoid triviality. I do not believe that this problem should be met by treating interesting general topics — and mycology abounds in these — within a framework of taxonomic groupings. However much one may deny it, students then gain the false impression that one's whole object is to teach life-cycles and systematics, and that a particular classification is either sacrosanct and permanent or at least better than others. The question, 'Better for what purpose?' is seldom raised.

As a practising taxonomist I do not decry taxonomy — far from it! But it is realistic to suppose that most students have little interest in a course that stresses formal classification, however unintentionally. Taxonomic knowledge comes by experience, not by the purgatory of memorizing a catalogue of names, diagnoses and keys. Most professional taxonomists are drawn into this field at a graduate level or later, by their own desire and efforts. The groundwork required is a broader view of fungi; the amount of classification given need be only enough to achieve precise communication and to connect topics. The balance is nevertheless difficult to strike. Seeing that science does not consist in gathering facts for their own sake, but rather in reasoning from observations, forming hypotheses and testing them, I believe that classification should be approached as the logical groupings arising out of comparative morphology and other studies; it should be the hypothesis which seeks to express the order to be found among the great diversity of fungal forms. This at least is the ideal even if in practice classification often becomes merely a catalogue of knowledge about fungal forms.

I consider that knowledge of the ways in which fungal classification may be built up is more important than a detailed survey of the taxa themselves. My object here has thus been to stress morphology and terminology in the first place — because they are basic to communication about fungi — and to use this knowledge at every appropriate stage to show how classification is arrived at. It is in this sense that I use the phrase 'principles of fungal taxonomy' as the title of this book, hoping that prospective readers will not be misled into expecting something more sophisticated. I have tried to introduce, and expand upon, a thread of classification throughout the book until the principal taxa are apparent and it then becomes clumsy and uneconomical to continue using this approach. At this stage it is inevitable that the taxa should be treated more formally, but not necessarily to the same level in each group studied. In general, the level selected and the examples studied serve the purpose of illustrating particular points to be made; for example, some taxonomic principle, biological diversity and versatility, or economic importance.

In stressing morphology, I realize that a descriptive catalogue of structures may be just as boring as one of fungal taxa unless it can be brought alive by relating structure to function and habitat. To do this at all adequately requires time and space, and a treatment of fungal physiology beyond the scope of this book; nevertheless, some attempt has been made to solve this problem. At first sight the emphasis on terminology may appear oppressive; but I consider it to be vital for precise communication and, as it is reinforced in practical periods, it has not, in fact, proved a serious obstacle to most students.

The general philosophy and principles of taxonomy are lively topics which can be relied upon to cause much interest and controversy among biologists. They should be introduced to students at an early stage, but are seldom mentioned. I thought it useful to give a brief account of some of these aspects (Chapters 2 and 3) before beginning to discuss fungi in detail. It is intended that these chapters should be browsed over originally, and perhaps re-read at a later stage.

I have not attempted to cite literature references for each statement made or section of the work covered, but have done so when necessary for developing an argument in the text. By doing this it may appear that I have not given credit where it is undoubtedly due. However, I am deeply aware that this book is a compilation in which my share has been to summarize, review or re-arrange the discoveries and thoughts of other mycologists. It is something like the method of Professor McGinty (Lloyd, 1913) who made all his notable discoveries and 'new combinations' seated at the top of a library ladder with a large book in his lap.

Most of the repetition which occurs in this text is deliberate, but I hope not excessive. A student, unlike a reviewer, does not read from cover to cover in a short time and thus needs to be reminded of certain facts which, although mentioned before, become appropriate again in a somewhat different context. I believe that this reinforcement of ideas and facts is necessary and also that it is helpful if the various chapters are more or less self-contained.

With a few minor changes, the broad classification adopted in this text is that of Ainsworth (1966).

While this book does not contain any practical exercises, it may be useful to state what I believe should be the aims of laboratory periods in an introductory course. These should not be primarily to train students to recognize and classify various types of fungi, although some knowledge of this kind will certainly be acquired incidentally. Practicals should instead be devoted to: observing and recording accurately what is seen; comparing fungi or parts of fungi in various ways, and writing down the comparisons; seeing the morphological structures mentioned in lectures and arousing curiosity about their functions; seeing living fungi in action; and reasoning from observations. These are the tools used in practical applications of mycology and are also the basis of taxonomy.

P.H.B.T.

Adelaide,
South Australia,
1969.

ACKNOWLEDGMENTS

I should like to thank my colleagues, Dr J. H. Warcup, Dr B. G. Clare and Mr J. A. Simpson, for kindly reading drafts of this book and making many helpful suggestions for its improvement; Mr B. A. Palk for photographing the illustrations; and Dr G. C. Ainsworth for his friendly encouragement.

Where I have quoted from publications or copied illustrations, the sources are cited in the text, but in addition I would like to record my thanks to the several authors and publishers who have kindly permitted the use of copyright material: The Rockefeller University Press for the use of a quotation from an article in the *Journal of Cell Biology* by Dr C. F. Robinow; Pergamon Press for material which appeared in *Soil Fungi and Soil Fertility* by Dr S. D. Garrett; Academic Press for a quotation from *Plant Diseases: Epidemics and Control* by Dr J. E. Van der Plank; The University of Minnesota Press, Minneapolis, for material which appeared in *The Molds and Man* by Dr C. M. Christensen; University of Chicago Press for material from an article in *Botanical Gazette* by Dr G. W. Martin; The International Association for Plant Taxonomy for a quotation from an article by Dr A. Munk which appeared in *Taxon*, and for extensive quotations from an article of my own in *Taxon*; The Editors of the *British Mycological Society Transactions* and *The Naturalist* for permitting quotations from articles by Mr E. W. Mason; The Systematics Association for material which appeared in an article by Dr A. D. J. Meeuse in the Association's Publication No. 6; The Colston Research Society for permitting me to quote from an article by Dr P. H. Gregory, and to copy an illustration by Dr M. F. Madelin, which appeared in *The Fungus Spore*; The New York Botanical Garden for allowing me to make a line drawing from a photograph by Dr D. I. Fennell and Dr J. H. Warcup which appeared in *Mycologia*.

I am much indebted to the Chief of the Botanical Research Institute, Pretoria, to Dr O. Vaartaja and to Dr Elaine Davison for allowing me to make line drawings from photographs in their possession.

CONTENTS

1

INTRODUCTION: ON THE NATURE AND IMPORTANCE OF FUNGI

The average person might associate fungi with moulds and mildews, mushrooms and toadstools, rusts and smuts, perhaps even with ringworm and athlete's foot, but usually he knows little about this large and varied group of living organisms. Robert Hooke, in 1665, was more inquiring and better informed when he wrote of fungi in *Micrographia*: 'The blue and white and several kinds of hairy mouldy spots . . . will not be unworthy of our more serious speculation and examination'; and again, '[They] consist of an infinite company of small filaments, every way context'd and woven together, so as to make a kind of cloth.' Mycology, or the study of fungi, deals with the structure and properties of living organisms, most of which are constructed in this way.

Traditionally, fungi are studied as non-vascular cryptogamic plants, which reproduce by means of spores, lack chlorophyll or other photosynthetic pigments, and have an assimilative body which may be amoeboid or unicellular in some species, but typically is made up of multicellular branching filaments (Fig. 1) called **hyphae** (s. **hypha**). In several respects fungi are plant-like: they are not motile except in amoeboid forms, but may have motile reproductive cells; their cells have definite cell walls agreeing in structure, but not in composition, with those of plants, and the cell contents are broadly similar; and they absorb nutrients through the cell wall, leaving residues outside. In contrast most animals are motile, ingest solid food and digest it in an alimentary canal (with some exceptions, notably in Protozoa), and lack cell walls. In their physiology, nutrition and behaviour, however fungi are less like plants. Green plants are **autotrophic** organisms, enabled by having chlorophyll to use radiant energy in synthesizing carbohydrates from simple inorganic materials. Fungi, however, are **heterotrophs** and in this respect are like animals: they lack chlorophyll and get their energy by using the organic matter of plants and animals as food. Oxygen is necessary for fungi and animals, but they do not assimilate carbon dioxide in the manner of green plants. Although some fungi are able to fix carbon dioxide chemically, often in organic acids, the process is not comparable with photosynthesis. The carbohydrates stored by fungi and animals are fats, glycogen and oils, while green plants store materials like fats, starch or cellulose. Other important storage products of fungi are the carbohydrate trehalose, and phosphate polymers such as volutin.

The more important elements required by fungi include: carbon, nitrogen, phosphorus, oxygen, hydrogen, potassium, magnesium and sulphur. Their vitamin and amino acid requirements are varied; some of these may be synthesized by the fungi themselves, together with a host of other metabolic products which are sometimes extracted for commercial use.

It is clear that the lack of chlorophyll in fungi must profoundly affect their construction and mode of life. Independent of light energy for photosynthesis, the fungus can occupy dark places and grow in any direction, invading the substratum with its microscopically fine hyphae. The absorptive hyphae generally remain almost prostrate, in close contact with sources of food and moisture in soil, decaying wood, living host tissues or other substrata containing organic material. Some fungi are able to absorb water and grow in soil whose moisture content is well below that of wilting point (Griffin, 1963), while some wood-rotting fungi obtain water by decomposition of cellulose in apparently dry wood. Despite their small size and apparent fragility, the hyphae are able to excrete enzymes and digest relatively hard organic substances in their path, then to absorb the digested material in solution. The enzymes produced by fungi partly govern what they are able to live on, and much of the development in fungi has been along the lines of physiological specialization. As heterotrophs requiring ready-made organic materials to break down for sources of energy, fungi may live either as **saprobes** on dead plants or animals, or as **parasites** on living plants or animals. They may be **obligate** parasites able to live, as far as we know, only as parasites. If their nutritional range is narrow, or other factors limit them to one or a few host species, they are sometimes termed **host-specific.** Many fungi, however, are parasitic or saprobic according to circumstances and are then termed **facultative** parasites.

In nature, organic and inorganic molecules are constantly being built up into more complex substances or broken down into simpler ones by the action of living organisms. These activities are well balanced. Together with other micro-organisms, the majority of fungi, as saprobes on dead organic matter, are indispensable in returning 'ashes to ashes and dust to dust', restoring soil fertility and preventing an accumulation of dead materials which would otherwise prohibit continuation of life on earth.

Fungi are very successful organisms, part of their success being due to their great plasticity and physiological versatility. They generally produce enormous numbers of spores which they liberate and disperse with great efficiency. Thus practically every ecological site — whether it be water, air, soil, a table top or the crook of one's elbow — not only has its fungal inhabitants but has staggeringly large numbers of them. It is the sheer ubiquity and number of these organisms, as well as their ability to live on, and change, a great variety of substrates that makes them so important in the every-day affairs of man. Even if fungi may not do particularly well in an adverse site they can often survive there until conditions change in their favour, or can make sufficient growth to enable them to produce a crop of spores, some of which may be dispersed to a better place. Even if their spores do not germinate and grow into new colonies, a mass of spores can of itself be a source of problems to man.

Many fungi are pathogens of plants grown for food, shelter and clothing, while a smaller number are agents of disease in animals, including man. Many saprobic fungi attack and degrade raw or manufactured materials of every kind, such as foodstuffs, timber, textiles, leather, paint, glue, plastics, chemicals, kerosene and diesel oil; even optical glass can be ruined by the presence of fungi which grow on dust particles or cementing-films in lenses and eventually etch the glass. The economic loss from such destruction, and the cost of preventive treatment, are enormous. Ramsbottom (1937) gives a fascinating account of the destruction caused by 'dry-rot' and other fungi to wooden ships of the Royal Navy in the 17th to 19th centuries. As a result of use of unseasoned timber and lack of ventilation, many of these ships became decayed by fungi and needed extensive repair soon after launching, while scores rotted at their moorings or foundered in service. Most notable was the *Queen Charlotte*, launched in 1810 at a cost of £88,534, which needed to be repaired even before going to sea; within 6 years the repair bill had amounted to more than her original cost. In those centuries of European wars, when sea power was vital, one can only conjecture how different the course of history might have been if the British fleet had not been laid low at times by the action of wood-rotting fungi.

To turn now to the other side of the picture, many saprobic fungi are useful to man. The use of yeasts in baking and brewing and as an industrial source of carbon dioxide and alcohol is well known. Food-yeast, a source of proteins, vitamins and amino acids, is manufactured by growing *Candida utilis* on cheap substrates such as potato starch waste, molasses, or sulphite liquor in the woodpulp and paper industry. Other notable products of yeast-like fungi are glutathione, used in the treatment of radiation sickness, and certain fats.

Minor sources of food include edible mushrooms, truffles, *Cyttaria* species, and edible sclerotia of some polypores such as *Polyporus mylittae*. Various fungi, notably *Penicillium roqueforti* and *P. camemberti*, are used to impart distinctive flavours to ripening cheeses. On the whole, however, fungi are not extensively used in western countries for fermenting raw agricultural products into foods. The position is different in many eastern countries where the staple diet, such as rice or soybean, is little varied and rather tasteless. Hesseltine (1965) has given a detailed account of some of the important foods made in Japan, China, Indonesia and the Philippines, by fermenting rice or soybean with various fungi, especially members of the Mucorales and Eurotiales, in order to give them flavour and make them more digestible. These fungi are among the first colonizers of raw substrates. The species used have the advantages of growing rapidly, producing abundant enzymes, being non-toxic as far as is known, and imparting a pleasant taste, odour and colour to the food. The Indonesian 'tempeh' is a food made by fermenting soybean with *Rhizopus oligosporus*; soybean alone is hard and indigestible but within 24 hours of treatment with this fungus it becomes softened and palatable, the proteins are broken down into water-soluble amino acids, and there is an increase in the content of riboflavin, niacin and vitamin B_{12}, but a decrease in thiamin. In Japan, 'koji' is produced from rice fermented by *Aspergillus oryzae* to hydrolyse the rice starch into simple sugars; the product is then mixed with soybean and salt and subjected to a second fermentation, this time with the yeast *Saccharomyces rouxii*, to produce a paste or soup known as 'miso'. Several other important foods, colouring agents, or flavourings such as soy-sauce, are made by fungal fermentations.

In contrast to fungi used in foods, some fungi are extremely poisonous, particularly those containing indole derivatives similar in constitution and action to the drug LSD, e.g. species of *Amanita* and *Psilocybe*, and *Claviceps purpurea*. Many species of *Amanita* are either dangerously or deadly poisonous. Some species of *Psilocybe*, which cause mental disturbance or hallucinations when eaten, have been used for centuries in the religious ceremonies of certain Mexican Indian tribes; psilocybin, an active principle in these mushrooms, has been synthesized for use in psychiatric research. The sclerotia of *Claviceps purpurea* (ergot) contain several alkaloids used medicinally in obstetrics and in the treatment of migraine and some vascular disorders. Fodder or grain supporting the growth of a mould, *Aspergillus flavus*, may be highly toxic to farm animals; the toxins concerned, called aflatoxins, are metabolic products of the mould and are active even in very small doses.

Among the fungal toxins affecting plants are some which are specific to the host plants of certain pathogenic fungi (Pringle and Scheffer, 1964). *Helminthosporium victoriae* produces such a toxin which is able to reduce the root growth of susceptible oat varieties by 50% at a dilution of over 1 : 1 million, but which has no effect on non-susceptible varieties even at high concentrations. The toxin is a polypeptide of relatively low molecular weight, and susceptibility to its action in oats is controlled by a single dominant gene of the host plant. The toxin appears to be responsible for producing all the symptoms of disease associated with the pathogen. Similar host-specific plant toxins are produced by *Periconia circinata* which causes the 'milo' disease of grain sorghum, and *Alternaria kikuchiana* which causes black-spot disease of Japanese pears.

Several of the extremely useful antibiotics are metabolic products of fungi, e.g. penicillin from *Penicillium notatum* and *P. chrysogenum*, and ceporin from species of *Cephalosporium*. Other potentially useful antibiotics are known: Beneke (1963) reported that calvacin, extracted from the Gasteromycete *Calvatia*, is effective against thirteen out of twenty-four types of cancer, but cannot be used clinically because severe allergies develop. Antibiotics are used extensively not only in clinical medicine but also in bacteriology and mycology in the preparation of selective media for isolating particular organisms from materials containing a mixture of organisms. Several organic acids, especially citric and gluconic acids, are manufactured commercially by fermentations using fungi. Fungi are also used in some stages of the conversion of sterol substrates to animal steroid hormones, the pharmaceutical use of which in contraceptives is of increasing importance. Such conversions are difficult to carry out by ordinary chemical means.

Fungi, as well as other micro-organisms, have recently become important as experimental organisms in many fields of biological research. Small organisms have a high ratio of surface area to volume or weight, and interaction with the surrounding medium thus takes place over a relatively large area. They absorb nutrients over their entire surface; thus comparatively large amounts are absorbed, and large amounts of metabolites are changed or rejected at their surface. The basic cell structures and life processes are shown by these organisms which can often be cultured easily, multiply quickly, occupy little space and are comparatively easy to observe and manipulate. They have been used, for example, in investigating metabolic pathways in biochemical processes; in studying growth phenomena; in elucidating mechanisms of cell division; in microbiological assays of vitamins, amino acids and trace elements; in studying processes of absorption on cell walls and the permeability of cell membranes; and in studying genetic and population phenomena.

The relationship of fungi to other living organisms is by no means easy to assess and a brief statement of prominent views on the question cannot hope to escape the accusation of being superficial. The reader is referred to Martin (1932, 1955, 1960) and Corner (1968) for stimulating discussions. Most fungi are relatively soft in texture and not readily preserved as fossils, while the few existing fossil fungi are not much help in phylogenetic study since they closely resemble modern forms. We can rely only on details of structure and life-cycle in modern examples and speculate on the ways in which these may have arisen in ancient times. There are closely similar structures in many modern fungi and algae which have led some mycologists to suggest that filamentous fungi were derived from filamentous algae, at various levels, by loss of photosynthetic ability and by adoption of the saprophytic or parasitic habit. The similarities with algae are most marked in some of the green and red filamentous algae, but it would be completely misleading to think of the divergence to fungi as having occurred among modern forms of algae or in the open sea. If fungi did originate from algae, they must have done so during the establishment of a land flora, from seaweeds among the rotting vegetation of lagoons and littoral swamps. Modern marine, freshwater and terrestrial fungi would all be derivative. The filamentous algae themselves were apparently derived polyphyletically from various types of flagellate plankton; thus fungi derived from algae would also be polyphyletic. An opposing view stresses that since modern fungi and algae differ radically in physiology and nutrition, development and cellular organization, and nuclear condition and behaviour, it is possible that the fungi were derived independently from colourless, flagellate stock. If so, it is postulated that many of the similarities in fungi and algae are due to parallel evolution in filamentous growth-forms, for filaments are restricted in their abilities and tend to produce a few characteristic types of structure under similar environmental conditions.

In this book I follow Ainsworth (1966) in treating the fungi (Mycota) as a Subkingdom of the Plant Kingdom. Two Divisions of fungi are recognized: the Myxomycota (slime-moulds) and the Eumycota. The chief difference between these is that in the Myxomycota the multinucleate protoplast is covered only by a plasma membrane and can move freely with an amoeboid movement in its assimilative phase. In Eumycota, the protoplast has varying degrees of ability to move within a system of hyphal walls, or within a single cell bounded by a wall; although the whole protoplast of the individual is multinucleate in the filamentous fungi, it usually becomes divided into compartments or 'cells' with a reduced number of nuclei.

2

SYSTEMATICS, TAXONOMY AND NOMENCLATURE

SYSTEMATIC STUDY OF FUNGI

How does one begin to study fungi without access to books and with no previous knowledge of fungal structure? Let us imagine the scene to be an uninhabited island where our hero, an observant and intelligent person of average background, has lately been stranded. He is hungry and notices that there are plenty of mushrooms nearby. He collects a large load of different types to take to the comfort of his rock shelter for sorting. He examines each specimen closely. Each has certain obvious features of shape and colour, smoothness or hairiness, and size, which give it individuality; yet each specimen can be matched with some others which appear to be essentially alike except for minor details, as far as he is able to judge from external appearances. He begins to sort the specimens into piles, ignoring small differences in the characters which he has almost unconsciously selected for special observation, and concentrating on features which are clearly different between one group and the next. He finds, for instance, that several specimens of about the same size and shape cannot be distinguished when viewed from above, but are easily distinguishable by the colour of the gills under the caps. This, he decides, is a useful differential character. The word 'species' is part of his vocabulary although he is a little vague as to its exact meaning; nevertheless, he congratulates himself on having sorted his mushrooms into 'species', one of which seems to match his memory of the common mushroom which he used to collect as a boy — the only type his family would eat. He keeps back a couple of specimens of this 'species' for future reference, and cooks and eats the rest without any ill effects. When he next collects mushrooms, he brings back only those which look like his reference specimens and he compares them carefully. But they are not quite the same; they turn bright yellow in places where the stem and cap have been handled, yet on the whole they are very similar to his edible 'species'. He is a little uneasy because he realizes that it is always possible to find small differences between any two individuals and that it is necessary to develop a sense of judgment on the degree of difference permissible between individuals placed in the same 'species'. Nevertheless, he is hungry and takes a chance. And at this stage we shall leave him, writhing in pain and reflecting that mushrooms are deceptive. (He recovered, by the way, and was rescued soon afterwards.)

This little allegory recapitulates some of the historical steps in fungal taxonomy: the desire to eat; collection of local specimens; the observation that some specimens are alike and others different in obvious features which can be seen with the naked eye; development of the idea that these are 'characters' by which fungi may be compared; development of some concept of a 'species' comprising individuals which may vary slightly in some of their characters but which are essentially alike and can be distinguished from other species by one or more clear-cut differences; the realization that some characters are less variable than others and that relatively stable characters are best to use for the purpose of differentiation; and finally, the realization that external appearances may be deceptive and that microscopic characters may be more stable.

To build up a knowledge of microscopic structures in fungi, and to be able to interpret them, would take an immense amount of time and insight, but eventually one might discover that specimens grouped according to similar external form (e.g. with cup-shaped fructifications) may sometimes be constructed of very different types of hyphae and may bear different types of reproductive cells and spores. The tasks of analysing fungal morphology would therefore be very great, and the difficulties of synthesizing the data into a meaningful classification, even greater. One's capabilities would have to rival those of Persoon, Fries, de Bary, Patouillard and some of the other giants of mycology: and one's life-span would need to exceed Methuselah's.

How different is the situation of the student with access to books on systematic mycology! He is shown how to recognize and interpret morphological structures seen in specimens. In determining species he is guided by the systematic keys into examining specimens in a certain order, establishing first those characters which differentiate the highest taxonomic categories (Divisions, Subdivisions), and subsequently looking for those which differentiate the lower categories in turn. In this way he soon gains an idea of which characters are considered to be important or definitive in each group of fungi, but will also realize that the importance of characters is relative and not of equal value in all groups. He will find that the definitive characters of the highest taxonomic categories are mostly microscopic, comprising especially features of the major types of reproductive cells and spores. At intermediate levels of classification the various forms of sporophores assume greater importance, while at generic and specific levels a whole host of criteria become important, particularly colour, shape, size and type of fructification, and septation and ornamentation of the spores. But he will also find that the weight attached to any one character at the species level may vary considerably from one group of fungi to another. Some species appear to differ only in a single character, but provided that this is constantly different it may suit one's purpose to recognize two distinct species.

So far we have spoken only of morphological criteria but it should be remembered that organisms have many other properties. All of these aspects must be studied before the true picture of fungal diversity is seen.

Finally there is the question of attaching names to the species and other categories of fungi that one is able to recognize. A little thought will show that unless there is some agreed method of naming, nothing but confusion can be expected. Such agreement is contained in the *International Code of Botanical Nomenclature*, the botanist's statute book. Botanists are allowed to differ in their ability to observe, and in the interpretations they put on their observations; those are skills and judgments which cannot be controlled. But attaching a name to the object of one's investigation and ranking it in a hierarchy of classification is subject to the laws set out in the Code. Laws may sometimes be changed, but the law as it stands must be obeyed.

RELATIONSHIP BETWEEN SYSTEMATICS, TAXONOMY AND NOMENCLATURE

Systematics, taxonomy and nomenclature are often defined differently in different texts. Taxonomy originally meant the principles underlying a system of classification; so presumably a taxonomist was one who studied and applied those principles. One would not suggest that taxonomists are unprincipled nowadays, but the concept of a taxonomist has broadened. He is employed to study, identify, name and classify fungi (or other organisms) and usually acts as the curator of a herbarium, with all that it implies. The best formula for expressing the relationship between these subjects would appear to be: Systematics = Taxonomy + Nomenclature (Lawrence, 1951; Bisby, 1953).

Because there are relatively few professional taxonomists it may not be realized that taxonomy is an everyday affair. All of us, consciously or subconsciously, study the things about us, make judgments on their properties and relationships, and classify not only the things themselves, but also our judgments about them. This is taxonomy, whether it concerns the mycologist studying fungi or the housewife comparing cheeses in a supermarket.

Two processes are concerned in taxonomy: analysis followed by synthesis. In taxonomic analysis the immense number of different kinds of organisms may be studied in a variety of ways (e.g. morphological, physiological, anatomical, genetical or ecological) in order to obtain as complete a picture as possible of the organisms with which one is working. The basis of taxonomic analysis is still mainly morphological, since structure is relatively easy to observe and record, and on the whole provides a good means of distinguishing organisms. More subtle differences, revealed by other fields of study (e.g. physiology) can be used in two ways: first they may be found to be reflected in the critical morphology of the organisms once the clue has been given that these may not be identical; second they can be used to distinguish between morphologically identical organisms if this suits the purpose of the taxonomist. The second process in taxonomy, namely synthesis, consists in meditating upon the data gained from analysis in order to classify them (and at the same time the organisms concerned), in a useful way. Taxonomy is therefore both the method and the practice of systematic classification.

Classification implies that the data obtained from analysis are rearranged in a systematic and useful manner. The best way to do this it to create taxonomic categories (**taxa**; s. **taxon**), each of which contains groups of organisms with common major properties. This stresses an often overlooked fact, to which attention was ably drawn by Luttrell (1958): that the taxa and their names are incidental to the wider task of classifying information about fungi.

Classification is not merely a system of grouping and ranking individuals in species, species in genera, genera in families, etc., and giving these groupings names for reference purposes; in addition, and more important, each name or taxon represents a series of generalizations which summarize what is known about groups of individuals (species), groups of species (genera), and so on. The taxa are arranged in a hierarchy of consecutive ranks which purport to reflect degrees of relationship between their constituent members. Among the lowest ranks in the hierarchy (e.g. species) the units comprising a particular taxon have very similar properties. Ascending to higher ranks in the hierarchy, the member-taxa are found to correspond in fewer and fewer characters, but those in which they do correspond are held to be of primary importance in reflecting a general relationship.

To be of use each taxon must be given a name for ease of communication and for recording information about it in an orderly manner. Naming, however, cannot be allowed to be haphazard; thus classification introduces nomenclature, which deals with the correct application of scientific names to taxa and the ranking of taxa in consecutive categories. The names given to taxa are only references — or heads of classification — under which data are grouped in an orderly and useful manner. Other heads of classification (e.g. ecological categories) could well be used, but they would not be nearly as numerous nor as useful for a wide range of purposes as the taxa that we use. Each name is a symbol for the information we have about a taxonomic group of fungi; each is tied to the mental picture we have of the characteristics held in common by members of that group, comprising perhaps a great diversity of fungal forms. Classification therefore becomes an hypothesis expressing the order to be found among the diversity of fungi.

Here, it should be observed that taxonomy is not dogmatic, as each taxonomist is entitled to his own opinion on the identity or otherwise of taxa, on the relative importance of various characters for the purpose of classification, and on the actual classification which he adopts. Nomenclature, on the other hand, is a rigid system allowing only a minimum of personal interpretation and decision.

Taxonomy is not primarily concerned with **identification**, which is merely the process of recognizing that two individuals are to all intents and purposes identical within the bounds of 'normal' variation. The assessment of variation is usually arbitrary. However, identification is achieved by taxonomic processes such as direct comparison of specimens and by the use of keys, descriptions and illustrations. Sometimes host indices, which list the species of fungi recorded on particular substrata, are useful in identification. No names need be concerned in identification: at a police identification parade one need only recognize the criminal; one need not name him or know anything else about him. Taxonomy is more concerned with **determination,** the naming and classification of organisms after suitable study. Determination therefore introduces nomenclature.

Taxonomy does not enjoy a high reputation among scientists[1] largely because it tends to be descriptive and non-experimental; indeed some would deny that it is a science, while nevertheless taking advantage of the results of its labours. Rogers (1958) has written very competently on the claims that taxonomy has to being considered scientific. Science is defined as an organized body of objective knowledge. The data that we get by various means are objective in the sense that there would be very great agreement between workers in the description and comparison of the same organisms. These data are then organized and generalized in the form of a classification; for classifications are the hypotheses of taxonomy. As with other hypotheses, we regard classifications as tentative, to be tested and modified in the light of new knowledge, and to provide a basis for prediction especially in applied fields of work. Thus taxonomy would appear to be completely scientific, but it fails in at least one respect: that its classifications are often utilitarian and can change not only with increase in knowledge but also with change in purpose. So classification is not entirely objective and is not universally accepted as being scientific.

As all new data must be integrated into the classification and change its pattern in varying degree, and as all organisms are subject to evolutionary change (though usually over a long period), it is virtually impossible to produce a completely stable classification. In consequence, the names given to

[1] Christensen (1961) speaks of 'those who spend their time putting living things into pigeonholes, a low but necessary form of scientific endeavour'.

taxa sometimes have to be changed, since the name reflects the classification; yet one of the chief aims of the nomenclatural system is to achieve fixity of names.

Only a completely phylogenetic classification would be stable and this is worth striving for as a matter of great practical importance no less than for the intellectual satisfaction it would give. A phylogenetic or evolutionary classification purports to show the genetic relationships among its various taxa; those taxa that are most closely related by ancestry are brought together, while unrelated taxa are widely separated. This grand evolutionary sequence was perhaps first visualized by l'Obel when he wrote in *Adversaria* of an 'order [i.e. a sequence], than which nothing more beautiful exists in the heavens, or in the mind of a wise man' (Arber, 1953). We may at times approach the ideal of a phylogenetic classification, but its complete realization is a pipe-dream since phylogeny can only be inferred from the similarities and differences in examples of fungi living today. We have very little factual knowledge of their ancestors.

A phylogenetic classification is often termed 'natural' in the sense that it is composed of 'natural' groups of organisms whose members agree in many features other than those actually used to group them. Views on what constitutes a 'natural' group in taxonomy have been discussed by Heslop-Harrison (1953) and Gilmour (1961). One view is that a natural group is one whose members are phylogenetically related, i.e. share a common ancestry; a second is that a natural group is one whose members share the largest number of common attributes. In effect these may amount to much the same thing: those groups sharing the largest number of common attributes can be considered to have resulted from evolution along a common line of descent. The empirical taxonomists are perhaps more honest in their approach to classification. They argue that one should not use speculation on phylogeny to establish taxonomic groups, and that a distinction must be made between a classification made possible by the influence of ancestry, and one based on phylogeny. Phylogenetic speculation is usually indulged in only after certain taxonomic groupings have already been made. These taxonomists are therefore unashamed at using an empirical approach and base their classifications on resemblances and differences existing in the materials at hand. Taking care to avoid analogy or the effects of convergent evolution, which can often produce a high degree of correspondence in characters of organisms that are in fact not closely related, it is then possible to arrive at a sound classification useful for the widest range of purposes.

Such a principle of classification lends itself to what has been called 'numerical taxonomy' and to the use of electronic data-processing machines for sorting organisms into groupings based on maximum similarity, which can then be made into taxonomic categories. There is no doubt of the usefulness of machines for storage and recovery of data, but the taximetric approach to classification is still largely at the experimental stage. Machines can process data extremely efficiently, but the taxonomic groupings which result from their use depend upon the questions asked of the machine, and must always be evaluated by reference back to specimens. The machine can give a number of alternative classifications (Kendrick and Weresub, 1966); some will be absurdly wide of the current groupings based on the computing power of the human brain; others will be close enough to be meaningful and to cause reappraisal of existing groupings and classifications.

In all taxonomy it is necessary to attempt to differentiate between **homologous** structures, which are fundamentally alike in their origin and development, and **analogous** or merely superficially similar structures. Corner (1959), stressing the point that it is misleading and also rather futile to compare structures or organisms which are not closely related, says that we should compare taxonomically related species of the same general ancestry; because they are closely related, but not identical, we can hope, by studying them, to obtain a picture of the course of evolution.

While one would not deny that observation of living fungi in the field and laboratory is essential, the proper place for assembling the body of systematic data on fungi is in the herbarium, in the form of specimens, records and publications. The herbarium should be, as defined by General J. C. Smuts in opening the National Herbarium in Pretoria, 'not only a collection of the plants of a country properly arranged and classified, but also the encyclopaedia, the catalogue, the storeroom and the laboratory of the botanist.' Alas, too often it is only a depressing storeroom laden with fumes of insect-repellents!

Importance of systematics

Systematic studies are important because they facilitate the naming and classification of organisms. Names are the only link between various fields of work, and the only means of intelligible communication. Since each unit of classification contains groups of organisms whose major properties are the same, we can predict much about newly discovered organisms once they have been classified; all advisory services depend upon this premise, and if the system should break down in practice it indicates a poor classification and points to further research that should be done. Systematic study is necessary in recognizing and sorting out mixtures of species which are commonly encountered in most applied work and often in biological research. It is axiomatic that one should know what one is working with, before beginning a piece of research or at least before publishing the results. Work done on materials which have not been sorted and named is utterly useless: the work cannot be repeated or gives conflicting results if the wrong name has been reported or if mixtures of species are present, not always in the same proportions. A knowledge of systematics is often essential in legislation and in litigation, in framing and administering any Act dealing with the eradication of noxious organisms, with the protection of rare plants, or with plant control and quarantine. Only by proper classification can records be kept, and found again when needed.

NOMENCLATURE

Nomenclature comes into account as soon as names are applied to species or higher or lower taxonomic categories. It deals with the laws and principles governing the correct application of scientific names to taxa, and also with the grouping of taxa into consecutive categories of definite rank, according to a particular nomenclatural system. The system we use is called the International Code of Botanical Nomenclature. It is subject to review and modification from time to time at International Botanical Congresses.

Nomenclature follows after taxonomy and decides the correct name to be applied to a species or other taxon within the particular classification adopted. Bisby (1953) remarks that nomenclature does not question taxonomic decisions as to the basis of a category, but merely seeks to give it its correct name *and* rank in accordance with the Code of Nomenclature. The practical sequence of events is first to make a taxonomic decision as to the categories in which a specimen must be classified, by identifying the specimen with a certain species and by placing that species in a genus; and second, to make a nomenclatural decision which gives the species its correct name within that genus and places the genus in higher taxa, i.e. gives it its correct rank.

Names of plants and fungi

Only after the year 1600 were plants grouped according to taxonomic affinities and not in relation to their use by man for food or medicinal purposes. In the earlier herbals the plants were catalogued under their local 'common names'; later short descriptive phrases were added, resulting in 'phrase names' in which the first one or two words had an almost generic significance and the remainder served to differentiate the species from similar ones. With hindsight, it is not too far a step from the descriptive phrase name to the binomial, in which the generic name and the specific epithet need have no reference to the characteristics of an organism, but are simply a code reference to its description in a formal publication.

One species may have several different names (**synonyms**), or several species may all have been given the same name (**homonyms**). This applies to both common and scientific names, but as the latter are codified one can check synonyms and homonyms in all cases and rule which is the one and only correct name to be applied to a species or other taxon.

A name is only a reference and becomes significant only when it is attached or related to the object of study. The name of a species is a nomenclatural entity, a part of language; it is not the same thing as the species itself. Similarly the **type specimen** is the type of a name, and not of a species; we use the type specimen to relate the name to representatives of the species. One can collect representatives of a species, but not the species itself; this usually persists despite our thoughtless attempts to exterminate it!

In English literature there are two opposing views on the purpose of names. Shakespeare says: 'What's in a name? A rose by any other name would smell as sweet.' But Lewis Carroll does not agree that a name is independent of the properties of the object that is named: 'Must a name mean something?' Alice asked doubtfully. 'Of course it must,' Humpty Dumpty said with a short laugh. 'My name means the shape I am.' Botanists are agreed that names need not be descriptive. It is perhaps desirable that they should commemorate some salient feature of the species, but as a shorthand reference they do not necessarily have to do so; all they have to do is to convey a single unambiguous idea.

The Code of Nomenclature gives guidance on the choice of names for new taxa and cautions against taking names from 'barbarous tongues', but tactfully omits to define these. To all but Polish mycologists it is no consolation that the names *Penicillium chrzaszczi* and *Pencicillium szulczewskii* have latinized endings, yet one would not call Polish a barbarous tongue.

Binomials

The species is named by a **specific epithet** placed after a **generic name** to form a binomial. In a formal citation the name of the author of the species may be placed after the binomial, e.g. *Thelephora cinerascens* Schweinitz. The names of authors are often abbreviated: e.g., Fr. (Fries); Pers. (Persoon); Schw. (Schweinitz); Berk. (Berkeley). But short names usually appear in full: e.g., Thom; Peck. The date of publication is sometimes added, and in a full **synonymy** there would be a complete reference to the journal in which the taxon was published. Citations of this sort ensure that there can be no confusion when two taxa have accidentally been given the same name by different authors.

Synonyms and homonyms

Two or more names considered to apply to the same taxon are called synonyms. **Obligate synonyms** are those names based on the same type material; **facultative synonyms** are names based on different type materials considered to belong to the same taxon. Homonymous names may arise as a result of ignorance that a particular name has already been published, or when a change in classification brings a species into a genus in which the specific epithet has already been preoccupied by another species. In the latter case a new name has to be proposed for the newcomer to the genus.

The principle of priority in nomenclature

Many synonymous names may be legitimate or valid, i.e. published in accordance with the rules laid down in the Code of Nomenclature, but only one of these can be the correct name for a taxon of given circumscription — in other words, the correct name for the taxon as classified by a particular author. Where authors disagree on taxonomy, one taxon may have several correct names under the Code; which of these is used depends upon one's taxonomic judgment. For example, *Corticium solani*, *Pellicularia filamentosa* and *Thanatephorus cucumeris* are all correct names for one taxon, although currently many taxonomists would probably favour the last name. In general, the earliest name legitimately applied to a particular taxon is the correct name for that taxon. For species, the correct name is a binomial composed of the earliest specific epithet combined with the name of the genus considered most appropriate in the classification adopted.

Name-changes

Although most names are relatively stable, they must be changed when for some good taxonomic reason the fungus is classed in a different category. *Thelephora cinerascens* Schw. was considered by Bresadola to be a species of *Lloydella* rather than one of *Thelephora*. The oldest specific epithet that Bresadola could find as applicable to the species was *cinerascens*, so he combined this with *Lloydella* to make the binomial *Lloydella cinerascens* (Schw.) Bresadola. The double citation of authors shows that the specific epithet is attributable to Schweinitz but that Bresadola had combined it in a genus different from that in which it was originally placed. Bresadola had made what is termed a **new combination.**

It is unfortunate that those groups of fungi which attract most research because of their practical importance, are likely to suffer frequent changes in nomenclature when new knowledge about them demands a change in their classification. This is a source of considerable annoyance to workers in applied fields, who are unused to thinking of species in terms of their synonymy.

Starting points for fungus nomenclature

To avoid the confused nomenclature of the early botanical writings, the Code states that names applied to plants before the publication of Linnaeus's *Species Plantarum* (1753) are not to be taken into account in establishing priority. This work is taken as the starting point for nomenclature of the Myxomycota. But Linnaeus, for all his eminence, had little knowledge of fungi, and even compiled some fungi in his section *Vermes* (worms) of his genus *Chaos*. Our knowledge has advanced and no one would wish to be still tied to that nomenclature. In mycology, therefore, the nomenclature of fungi begins with certain later authors: Persoon (1801), *Synopsis Methodica Fungorum* (for Uredinales, Ustilaginales and Gasteromycetes); Fries (1821–1832), *Systema Mycologicum* (including the *Elenchus Fungorum*, 1828) (for other fungi); Sternberg (1820); *Flora der Vorwelt*, Versuch 1 (for fossil fungi).

Names of fungi published before these dates are not to be taken into account unless they were taken into use in the above starting-point books; in that case they may be cited as though they had been described for the first time by Persoon, Fries or Sternberg. For example, *Agaricus melleus* Vahl, accepted by Fries in the *Systema*, may be cited as *Agaricus melleus* Fries; but it is more often cited as *Agaricus melleus* Vahl ex Fries to indicate its earlier history. In its modern classification this species is called *Armillariella mellea* (Vahl ex Fries) Karsten.

Names of fungi with a pleomorphic life-cycle

Later, it will be shown that many fungi have more than one morphological state, producing different types of spores in the life-cycle, and that these states are often extremely difficult to link with one another. Such different states may have been found and named independently without the realization that they belonged to only one species of fungus.

An important pathogen of potato and many other crops usually occurs in a distinctive mycelial, non-sporing state, which Kühn (1858) called *Rhizoctonia solani* Kühn. What appeared to be a different species — because it attacked cucumber and produced a fructification with basidiospores — was found by Frank (1883) and named *Hypochnus cucumeris* Frank. In 1956, Donk decided that these were simply different states in the life-cycle of a single species and, for taxonomic reasons, transferred Frank's species to the genus *Thanatephorus*, making the new combination *Thanatephorus cucumeris* (Frank) Donk. For those who accept Donk's taxonomy the official name of this fungus is *Thanatephorus cucumeris* because this refers to the sexual or 'perfect' state of the fungus. But it is permissible to refer to *Rhizoctonia solani* when the fungus is found occurring in the asexual or 'imperfect' state. Thus in mycology a single species may legitimately be known by more than one name according to the morphological state in which it is found. The Code rules that 'in Ascomycetes and Basidiomycetes with two or more states in the life-cycle, but not in Phycomycetes, the first legitimate epithet applied to the perfect state takes precedence'.

Many mycologists, endeavouring to recognize only one name for a species, will refer to a species found in its imperfect state by the name given to its perfect state, if this connexion is known. But there is a danger in indiscriminate application of this procedure, as often occurs in reviewing journals. Many sterile mycelia of some economic importance are given specific names despite the fact that they exhibit so few really distinctive taxonomic characters. Species of the genus *Rhizoctonia* are a good example: their mycelia may appear to be distinctive, but several isolates placed by competent taxonomists in a single species may form quite different perfect states if the trouble is taken to induce them to produce fructifications (Warcup and Talbot, 1966).

Taxonomic groups (Taxa)

The following suffixes attached to the stem of the scientific name of a **taxon,** indicate the various ranks in the hierarchy of classification:

Kingdom	Plantae
Division	**-mycota**
Subdivision	**-mycotina**
Class	**-mycetes**
Subclass	**-mycetidae**
Order	**-ales**
Family	**-aceae**
Genus, species and subspecific taxa	no standard suffixes

Unorthodox nomenclature

Most mycologists try to abide by the Code of Nomenclature but lapses do sometimes occur. Perhaps the most entertaining of these were perpetrated deliberately by Curtis G. Lloyd of Cincinnati, Ohio, in his seven volumes of *Mycological Writings* (1898–1925). Lloyd poked fun, with serious intent, at the practice of 'name-juggling' — undoing the work of someone else in order to gratify personal vanity by citing one's own name as the author of a new combination. Lloyd would not cite authors of names but reserved them for the 'Index and Personal Advertisements' section of an 'Advertising Supplement'. However, this attitude posed a problem for Lloyd himself, since he was the author of an exceedingly large number of new taxa and new combinations. So he invented the mythical Professor N. J. McGinty of Pumpkinville Polymorphic Institute and facetiously made new combinations in the name of McGinty. Names proposed by McGinty are held to be provisional names — *nomina provisioria* — and are formally cited as 'Lloyd as McGinty'.

Another of Lloyd's engaging habits was to draw attention to 'Bulls' or 'Synonyms, Mistakes and Blunders' published by his predecessors and contemporaries. These were always published with a portrait of Tso Kay, a little idol with an enigmatic smile, of whom Lloyd said: 'Behind that cheery countenance is just a little hint of hypocrisy that symbolizes much of the work done in mycology.' This is a harsh judgment. But despite his continual campaign against 'name-juggling,' Lloyd recognized that 'you cannot stop bullfighting by appealing to the matadors.' Who wants to stop the bullfighting, anyway? Taxonomy and nomenclature are not static!

3

TAXONOMIC PROBLEMS ASSOCIATED WITH VARIATION IN FUNGI

The reader is advised that although this chapter is a continuation of the preceding one discussing taxonomy and nomenclature in general terms, it assumes a knowledge of some of the structures and properties of fungi which are dealt with only in later chapters. It may therefore be necessary to return to this chapter at a stage when its content may be better appreciated.

The effect of variation in causing taxonomic problems is an important theme which seldom gets the attention it deserves, even from practising taxonomists.

At present taxonomy is largely concerned with determining species, and our emphasis is on species as the unit of classification. We should have less difficulty in determining species if species were definable in terms upon which everyone could agree, and if some fungi did not vary so much as to upset our judgment and agreement on the limits of a species. But because fungi vary, species are essentially unstable and are defined in various ways according to one's interest in different aspects of their variation. Some people are interested in the morphological, others in the genetic variation of fungi, and each has different ideas on what constitutes a species. Thus species are fairly arbitrary units, not equivalent from one group of fungi to another even when they are defined in the same terms. For example, with species defined morphologically, differences may rest on single obvious characters or upon a multiplicity of characters.

All organisms have a latent capacity for variation, which becomes detectable when morphogenesis is affected by changes in various internal and external factors. Broadly speaking, the variations are temporary, reversible and non-inheritable when the environment is altered, but usually permanent and inheritable when the genetic constitution is altered. The fungus as we see it (the **phenotype**) is the product of interaction between its genetic constitution and its environment (Burges, 1955). Regardless of the causes or mechanisms of variation the effects produced in the phenotype may be much the same: they are expressed as major or minor changes in its morphology and physiology.

It is not my purpose to deal here at any length with the types, causes or mechanisms of variation, but instead to stress its practical effects in producing taxonomic problems. First, change during the normal course of development of a fungus changes the taxonomic picture. There may, for example, be great difficulty in matching immature or even mature fructifications in culture with mature ones taken from nature. External factors such as light, **pH**, humidity and nutrition are often exaggerated in culture, and fungi can survive marked changes in these, but often respond with greater phenotypic variation. Such changes are generally temporary and reversible; but this cannot be said of a variant which is killed and made into a herbarium specimen. It has already been observed that pleomorphism (a type of developmental variation) has a direct effect on nomenclature, making a dual or multiple system of nomenclature a necessity. Finally, various types of genetic variation exist, e.g. those caused by hybridization and genetic recombination, by heterocaryosis and by mutation. Their effects are permanent and the variants are easily perpetuated by vegetative growth from cells with an altered nuclear complement.

Sterile mycelia are much alike in many features and there is usually more hope of identification if they are induced to form fructifications. But this is sometimes only the beginning of further difficulties. The precise conditions required for fruiting are seldom known and thus the cultural environment is usually manipulated in the hope of getting fruitbodies; sometimes this is successful, but the desired fructifications may vary widely, one from another, and from those of described species. One can sometimes produce a series of fructifications with characters varying continuously between those of two described species, thus suggesting that one is dealing with an **aggregate species** consisting of a number of different genotypes with the same general phenotypic resemblance at a given point in time. Where morphology fails to give a satisfactory answer as to the identity of isolates, other criteria can be called in. These may be genetical, such as the use of interfertility tests, or physiological, such as a comparison of protein and enzyme patterns produced by gel electrophoresis; but they can seldom give an unequivocal answer on the question of identity. The decision is quite as arbitrary as when it is made on a morphological basis. I shall return later to the question of interfertility tests but meanwhile I wish to stress that the units involved do not necessarily coincide with taxonomic species, since species concepts differ according to the particular aspects of variation which interest the investigator.

SPECIES CONCEPTS

The question, 'What is a species?', is meaningless without qualification and so is rightly evaded by most biologists when asked in that form.

The only inclusive definition of a species is perhaps that given by Meeuse (1964): 'A species is a group of individuals to which a binomial is given.' But as Meeuse pointed out this is an unreal and unhelpful definition. It defines a **nomenclatural species,** which Mason (1940) called an 'ink-and-paper affair . . . anchored to reality only by the designation of a type specimen'. It speaks of individuals, but there is controversy as to whether species actually occur in nature. Because biologists study individuals from natural populations, and for their own purposes arrange these in groups which they call species, they tend to assume that equivalent species occur in nature. Golovin (1958) says (translated): 'The species is a main unit, actually existing in nature, stabilized by heredity and representing a stage in evolution.' The opposite is clearly stated by Hochreutiner (1929): 'In nature there are no families, no genera, no species. . . . There are only individuals more or less resembling one another.' How can these contrary views be reconciled? Clearly it depends upon the kind of species one has in mind; thus it is sometimes necessary to review species concepts. These can be fitted into two main categories: taxonomic species and natural or biological species.

In nature, a group of individuals, presumably of common origin but themselves not identical, constitutes a population whose members are generally able to inbreed. The different populations constituting **natural** or **biological species** are separated from one another by a barrier to, or lack of capacity for, interbreeding. There can be little doubt that such species do exist in nature as more or less definite groups, and that they are subject to evolutionary change. On the other hand, some individuals occurring in nature resemble one another closely, or differ, in ways which the taxonomist has found easy to investigate and describe, and which he regards as important because he can use them to create groupings which suit practical purposes in most cases. These groups are the **taxonomic species.** They are as permanent as the written record but may have little reality apart from the taxonomist's system of classification.

Man-made taxonomic species and natural biological species may approach each other closely; but we cannot assume that they are equivalent, and can sometimes show that they are not. The taxonomic species is used mainly by morphologists and physiologists; the biological species mainly by geneticists and population biologists. In fact, however, the biological species is an ideal belonging to phylogenetic theory rather than a practical unit. It is probably true to say that no biological species has ever been properly delimited and described in practical taxonomy, and that the task of building a complete classification of biological species appears insuperable. Biological species always emerge only as segregates within the existing framework of taxonomic species.

Taxonomic species

Taxonomic species are determined by normal taxonomic methods which are usually morphological or physiological. Morphology is used to compare and group individuals, but it may fail when there is little morphological diversity, or when the groupings do not reflect physiological differences of practical importance. Then, physiological criteria are called in to assist in determining the species; these may be differential fermentations, temperature optima, host specificity of pathogens (often only presumptive), or electrophoretic patterns of proteins and enzymes, to mention just a few possibilities.

Taxonomic species are not equivalent from one group of fungi to another; they may depend upon single or multiple differences which may or may not be investigated. This would appear to depend upon the whim of the taxonomist, but in fact it depends upon his experience and purpose, his recognition of the extent of variation and the emphasis which he places on its degree. In circumscribing a species he looks for features which are apparently unique or at least do not intergrade with those of other species. Assessment of variation is, however, very difficult. Any biochemical or biophysical tests may be helpful: if differences are found they suggest that the morphology of the fungi should be checked again, or that one is dealing with an aggregate species. With any method of analysis, however, one must finally take an arbitrary decision as to whether two organisms which are not identical should be placed in the same unit of classification. The decision is still arbitrary when the data are processed by computers.

Taximetrists usually congratulate themselves on an objective comparison of a large number of characters in different organisms, but tend to forget that the characters are a chosen fraction of the total characters of the organisms being compared. Kendrick and Weresub (1966) showed that at the ordinal level it is extremely difficult to find a large enough number of characters to subject to computer analysis, and many characters which they used would normally be regarded as perhaps only of specific or generic importance in conservative taxonomy. Another thing often forgotten by taximetrists is that a large number of characters may collectively make one cardinal character which the conservative taxonomist seizes upon and interprets in a flash of old-fashioned brain-computing, without breaking it down into its components. His brain computes a particular grouping of characters as '*Erysiphe*'; it takes for granted that a cleistothecium contains asci, that the asci contain ascospores, and that the cleistothecium is a closed, rounded ascocarp and has a peridium. These are all associated characters; they do not exist independently, yet the taximetrist scores them separately and considers that he is achieving greater objectivity by using more characters. Even more important, perhaps, is the fact that a species itself cannot be defined objectively. Meeuse (1964) says: 'This is one of the principal reasons why the "numerical" approach has at least as much *a priori* bias as traditional methods of classification. A subsequent exact mathematical analysis ... does not eliminate the initial subjective element, so that the claim of greater exactitude is utterly unfounded and even misleading.... These methods substitute classifications without necessarily producing ones that are taxonomically sounder.'

The type system

A type specimen typifies a name and not a species; we merely hope that it is representative enough to form a standard of comparison through which the name may be related to the species. But the type system has grave disadvantages. The type specimen usually has the merit of being a random sample from a living population, but can never fully represent a species. Taxonomists know that the taxonomic species cannot cope with a continuum of variation such as is found in populations, but they are practical enough to know that one cannot use a continually changing standard, and that one needs a static type and a static concept of a particular species which can be expressed in writing. In practice this often works well, especially when species are separated by clear discontinuities. But discontinuity is often only assumed and two herbarium 'species' may merely be the extremes of a graded series whose intermediates are either not found, or found only when specimens are accumulated from many areas. Since the type is a phenotype, not a genotype, it is related to the population only at the time of collection. Then it is dried or killed, and buried in a herbarium. Except in rare instances when spores or other parts have survived this treatment, the type cannot be brought to life again and grown for comparison with living cultures under the same environmental conditions. Although the population is dynamic and subject to continual change, phenotypes drawn from it at a later date still have to be compared with the dead type specimens or with descriptions based on them. With populations in nature which are unlikely to have evolved very much in the relatively short period of recorded taxonomy, this may be of theoretical importance only. On the other hand, the generation time of such fungi as moulds is extraordinarily short both in nature and in culture. Their mutation rate might well be high; it can doubtless be measured, but its taxonomic significance often cannot.

Cultures

Often, much difficulty is experienced when the taxonomist works with experimental biologists who tend to use cultures rather than specimens taken from nature. Fruitbodies formed in culture may sometimes differ greatly from those of the same species in nature. The great difficulty is that type specimens, descriptions, and our whole systems of classification have been based mainly on natural fruitbodies. Synnemata of natural fructifications may become sporodochia in culture; sporodochia may become discrete sporophores; and an agar culture cannot show one whether the pycnidia are immersed or erumpent on natural substrata. Thus accepted keys and classifications continually break down with some fungi in culture.

Repeated subculturing at short intervals can result in a short generation time with a theoretically greater chance of genetic variation occurring. Even a hot inoculation needle can induce mutation (Barnes, 1928, 1930). Single-spore cultures are often favoured because of their genetic purity, but as the nuclear composition and nuclear behaviour of the fungus are seldom investigated first, one does not usually know the nuclear properties of the single spores; by using such cultures one may merely be segregating the organism into various genetic types and perpetuating only some of these in culture. It would seem axiomatic that standard media should be used in comparing isolates or species in culture, but it is often overlooked that two species may appear identical on the chosen standard medium, but differ precisely because they respond differently on other media or under other conditions. Therefore, a range of media and environments may be necessary.

Many of the older mycologists were contemptuous of culture work because they recognized that such methods may endanger the conventional classifications based on specimens taken from nature. Others have been sympathetic to culture methods, but have clearly pointed out their dangers. Mason (1940), to who e owe much for his cautious and balanced use of both natural specimens and cultures, speaking of the study of some fungi in culture, said: 'Their study is not the study of plants as they occur, but the study of plants as they do not occur. The game is up if they are not named before they are planted in a test tube'; and again (Mason, 1948) 'Their form [morphology] need not be established in order, so that they can be taxonomised into species; it is their pedigree which must be established . . .'. These statements show clearly the clash between species based on phenotypes and on genotypes; they also show that a nice balance must be struck between the study of fungi from nature and in culture. They are complementary studies; neither by itself gives a complete or undistorted picture of the fungus. Mason also pointed out that one tends to study individual organs and not an individual fungus in all its phases, and that one cannot establish the full range of variation of a species simply by taking a single isolate and multiplying it repeatedly in culture.

Biological species

The realization that wide morphological variation can exist in organisms of the same genetic stock has caused a shift in emphasis from studies of individuals to populations, and from a taxonomic to a biological concept of species.

The idea of interfertility between members of a species was first stated by Lindley in 1832 (quoted in Heslop-Harrison, 1953). Several species definitions are based on this idea, and a fair example is that of Mayr (Mayr, 1940: quoted in Heslop-Harrison, 1953): 'Species are actually or potentially interbreeding populations which are reproductively isolated from other such groups.'

There are several practical objections to interfertility as a criterion of species. The first is the practical aspect of describing a biological species in terms which make it realistic to other taxonomists. If one describes a particular species as including any population which is interfertile with certain culture kept at the Centraalbureau voor Schimmelcultures at Baarn, this does not help others to visualize a morphological entity; yet our whole classical system of recognizing species is founded on morphology which can be described.

Another practical objection is that interfertility tests are laborious and invariably done in culture. Much of this work has been done with Basidiomycotina, particularly members of the Polyporaceae. Dicaryotic mycelia obtained by culturing tissues of fruitbodies, and monocaryotic mycelia grown from isolated single spores, can be used. At its simplest, an interfertility test consists in opposing pairs of inocula on agar and observing whether the hyphae of the opposed mycelia are able to anastomose. If clamp connexions are known to be present in the dicaryotic mycelium of the species under investigation, further evidence of interfertility can be gained by opposing pairs of monocaryotic inocula (which lack clamps) and noting the formation of clamp connexions after anastomosis of compatible mycelia. Third, the 'Buller Phenomenon' can be used as evidence of interfertility: this is based on the fact that a large monocaryotic mycelium (lacking clamps) can be dicaryotized, and quickly develops clamps throughout its hyphae, if a small inoculum of dicaryotic mycelium of the same species is placed near its periphery and anastomoses with it.

The absence of appropriate compatibility factors does not prevent anastomosis from occurring, at least in some instances, but their presence is essential for clamp-formation in the anastomosed hyphae. Despite this some mycologists maintain that a positive interfertility test, as judged by the presence of anastomosis or of clamp-formation between opposed isolates, gives conclusive proof of specific identity. With Basidiomycotina, however, the presumably dicaryotized mycelia are seldom taken on to form fruitbodies and there is usually no assurance that they are able to do so and produce viable spores. Moreover, monocaryotic mycelia are sometimes able to form fruitbodies. Anastomosis has been shown by Bourchier (1957) to depend not only upon genetic compatibility but also upon such factors as temperature and nutrition. Flentje and Stretton (1964) have shown with *Rhizoctonia solani* that apparently successful anastomosis may be followed by a killing reaction involving several cells at, and adjacent to, the anastomosed cells; and McKenzie (unpublished thesis) showed that only at high temperatures could some of the dicaryotized cells grow away fast enough to escape killing and establish a mycelium. Various workers in Canada have given examples of degrees of interfertility between populations, of inconstant pairing reactions, and of the formation of illegitimate or unstable dicaryons which later become 'de-dicaryotized'. One must conclude that dicaryotization is only presumptive evidence, but not conclusive proof, that opposed isolates belong to one biological species but not necessarily to one taxonomic species. From a negative interfertility test no conclusions can be drawn; some simple physiological barrier (e.g. temperature) may prevent interfertility, or a different biological species may be concerned.

Again, it should be stressed that the units of orthodox classification, taxonomic species, may be close to, but should not be equated with, units defined on the basis of interfertility. The great dilemma in taxonomy is to fit variable organisms into stable taxa, and no combination of criteria can give unequivocal separation into constant groupings. Moreover the purpose of the classification is often an overriding factor. All attempts to classify organisms suffer from the same disability — where to draw the line. According to Van der Plank (1963) even Nature herself 'seldom draws lines without smudging them'. Munk (1962), quoting a verse by Piet Hein, has put it another way; he blames us rather than Nature:

'Our simple problems often grew
To mysteries, we fumbled over,
Because of lines we nimbly drew
And later neatly stumbled over.'

Despite their imperfections the conventional classifications are nevertheless of proven usefulness. They are useful for a wide range of purposes, cover all known organisms, and would be extremely difficult to replace.

4
MORPHOLOGY OF SOMATIC STRUCTURES

THE THALLUS

The assimilative body, or soma, of a fungus is a **thallus** (p. **thalli**), a comparatively simple growth-form lacking differentiation into true stems, roots or leaves. Despite this apparent simplicity, fungi show great diversity in size, metabolic activity and organization of their characteristic fructifications.

The thallus is an amoeboid **plasmodium** (Fig. 23) and lacks a true cell wall (having only a hyaloplasm covering) in the slime-moulds (Myxomycota); it is unicellular with a true wall in some representatives of the other Division of fungi (Eumycota), but typically in Eumycota it is composed of hyphae. Loosely aggregated somatic hyphae are known collectively as a **mycelium** (p. **mycelia**). When unicellular thalli produce bud-cells in succession these may remain attached to one another in an easily dissociated chain known as a **pseudomycelium**; the bud-cells may remain small or may become elongated and irregular in size. As more than one bud may form from an existing cell, the chains are often branched. The occurrence of a pseudomycelium is especially notable in yeast-like fungi, but some normally filamentous types (e.g. Mucorales) may take a pseudomycelial form in the presence of a high sugar concentration. Thus one may speak of a plasmodial, unicellular, pseudomycelial, or mycelial thallus.

In fungi which are internal parasites of plants or animals the thallus may be much reduced in size and even confined to a single host cell. Species with **dimorphic** thalli are not uncommon among such fungi; for example, some of the fungal pathogens of animals are unicellular and yeast-like in the host, but mycelial in culture, whereas certain plant pathogens, such as the leaf-curl fungus *Taphrina* and members of the Ustilaginales, have a mycelial thallus in the host but a yeast-like thallus in culture. The thalli of fungi thus range from ephemeral, single cells occupying small sites, to massive, multicellular perennial growths such as those of the large wood-rotting bracket-fungi, or those of the fairy-ring mushrooms, whose assimilative parts occupy a large area and may persist for decades in wood or soil, respectively, but are seldom noticed.

Eventually, in the normal course of events, the thallus gives rise to reproductive structures. Some simple types of thalli are converted entirely into reproductive cells and are termed **holocarpic**; in **eucarpic** thalli, which are typical of most fungi, only part of the thallus becomes reproductive and the cells formed for this purpose are usually highly specialized in form and development. In contrast, some types of survival structures formed from the thallus are relatively undifferentiated and thus tend to be widely distributed among different groups of fungi (see e.g. chlamydospores, p. 136).

Most 'specimens' in herbaria are simply the conspicuous fruitbodies or smaller spore-bearing parts of fungi, and the thallus itself is usually overlooked. One may object that it is impossible to make a herbarium specimen of a tree trunk or several square metres of soil in order to preserve the thallus. Nevertheless, it would be possible to sample these and above all to study the samples for taxonomic purposes; unfortunately this is seldom done.

Particularly in reference to fungi in culture, one often speaks of fungal **colonies**. A colony of yeasts comprises a mass of individual thalli living together. With mycelial fungi the colony is a mass of hyphae, often with spores, and, depending on how it originated it may contain one or more individual thalli. With both yeasts and mycelial fungi the colony is generally, but not always, of one species.

Much of this text will be concerned with the diversity of fungi and with differences and similarities in morphology which assist us in grouping them and formulating classifications. Nevertheless, the majority of fungi have a mycelial thallus and an essential unity in construction resulting from their being composed of hyphae which have certain capabilities and limitations.

THE HYPHA

A hypha is a microscopic fungus filament, usually branched, composed of an outer wall and a cavity (**lumen**) lined or filled with protoplasm. A new hypha originates by germination of some kind of propagule, usually a **spore**, which puts out one or more bud-like processes (**germtubes**) which in turn elongate and become hyphae (Fig. 1). There is no clear demarcation between a germtube and a young hypha: the one merges imperceptibly into the other. At intervals the hypha becomes divided by few or many transverse walls (**septa**; s. **septum**) into compartments known as **cells**, but since the septa are formed by centripetal growth of wall material and are initially and in many fungi permanently perforated at the centre, such cells often communicate directly with one another, in which case the cytoplasm and other cell

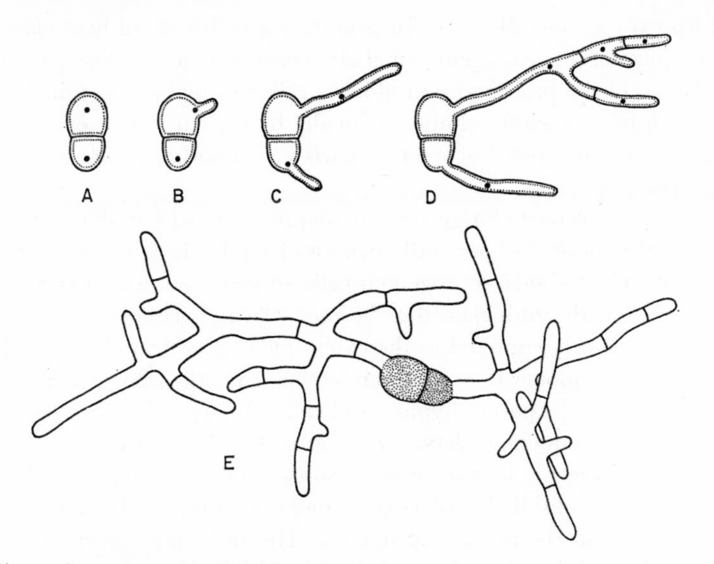

Fig. 1. Stages in germination of a two-celled spore (semi-diagrammatic). A, spore; B, C, germtubes; D, E, hyphae.

contents may be capable of streaming from cell to cell. In some fungi (e.g. Myxomycota, Mastigomycotina, Zygomycotina) cytoplasmic flow is often conspicuous in the thallus or in somatic hyphae. In others it may not be obvious until reproductive parts are about to be formed, but then the cytoplasm flows towards these parts and may leave some hyphae highly vacuolated and often virtually empty.

Apical growth of hyphae

The apex of a hypha is a thin-walled plastic region where growth materials are added, differentiation takes place, elongation occurs in a zone behind the tip, and where width and shape of the hypha are able to vary. Further back from the apex the wall sets and becomes rigid in time, and at least one more layer of wall substance is added. Hyphae absorb nutrients in solution through their walls; solids are digested externally by enzymes secreted through the walls and are then absorbed in solution. This process must set up an internal pressure on the more rigid parts of the wall, which is communicated to the plastic apex and probably assists in its forward elongation. As Robertson (1965) has shown, apical (**distal**) growth is the key to most fungal morphology. By interference with the balance between extension and setting of the wall, the apex can be made to expand or constrict, gradually or abruptly. Thus the hypha is not necessarily a cylindrical tube. In spaces between soil particles hyphae will often expand to fill a space or constrict to negotiate a narrow passage and may consequently become highly irregular in width over a short distance. In penetrating cell walls of host plants, the hyphal apices of the pathogenic fungi often narrow to a very fine 'penetration peg'. Many of the basic spore-producing cells in fungi are terminal hyphal cells which may be either characteristically blown out as sacs or vesicles, or constricted into narrowed or spicular parts which both generate and support the spores.

The more rigid part of a growing hypha may increase in diameter and in thickness of its wall, but generally cannot elongate. It can in some circumstances give rise to buds, sporogenous cells, or branches, which suggests that small areas remain unthickened or become plastic once more; branching is thus not necessarily confined to the extensible region near the apex though it is possibly commoner there. Branch hyphae have the same general growth characteristics as the parent hyphae and can in turn produce branches of the second and subsequent orders. Distal growth of the primary or 'leader' hyphae causes the whole margin of a thallus to advance centrifugally, while the presence of subsidiary orders of branching increases the density of the mycelium behind the advancing margin. The radiating leader hyphae are usually well spaced apart, while the **proximal** branches, towards the centre of the thallus, become closely interwoven as a mat — the 'kind of cloth' of which Robert Hooke wrote. Even in a simple thallus there is usually some later differentiation in function of some of the hyphae or their branchlets, which may be accompanied by marked morphological change. Some obvious examples of specialized somatic structures are considered later in this chapter. However, differences in hyphae are not always obvious even though present (e.g. in aerial as opposed to submerged hyphae), and this may pose problems in exact description for taxonomic purposes.

A major property of the hyphae of higher fungi is the ability to **anastomose** (Fig. 2). Neighbouring hyphae in the thallus are stimulated to put out short branches which make contact, or the tips of hyphae may make contact; the walls are dissolved at the point of contact and the result is a short continuous tube joining the two hyphae. The presence of abundant anastomoses creates a strong mycelial network and possibly provides for rapid and efficient cytoplasmic flow, co-ordinated growth of tissues, and for exchange of nuclei and genetic materials. Anastomosis is also possible between different thalli of the same species. When this occurs in a limited space it must result in a pooling of resources and the greater likelihood of a well nourished thallus for the production of fructifications.

Effects of centrifugal growth on the thallus

Since growth of the thallus is centrifugal, the younger and more active parts are towards the periphery. Some, at least, of the proximal hyphae tend to become progressively more vacuolated with age, giving up cytoplasm in a flow to younger branches and finally losing most of their contents and perhaps

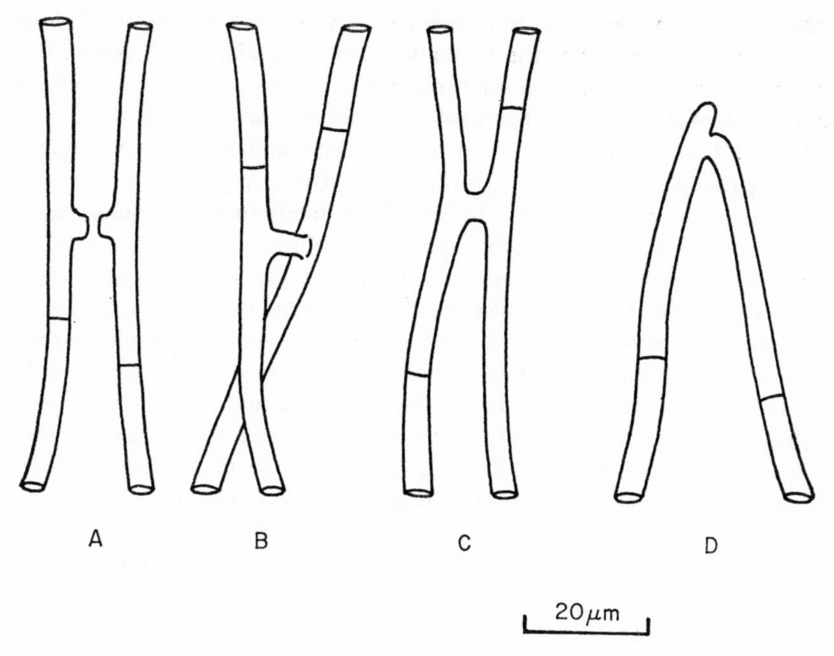

Fig. 2. Anastomosis of hyphae in *Corticium atrovirens*.

dying. The mechanism of cytoplasmic flow is still not well understood but may possibly be due to pressure from expanding vacuoles. The dead hyphae, composed simply of wall material which is usually surprisingly rigid, still play a part in supporting the living parts of the thallus on its substratum and probably also in holding water in their meshes like a sponge. For these reasons, mycelial thalli on a hard support spread more or less in a circle. In species which have small spore-bearing structures (**sporophores**) and take only a few hours or days to form them, such as most moulds in culture, any part of the thallus may be active enough to give rise to sporophores and spores. Many fungi, however, produce massive compound sporophores (**fruitbodies**), which are often seasonal in occurrence, and for which materials must be accumulated over a long period. In soil-inhabiting species of this type, e.g. the common mushroom, the fruitbodies tend to arise in a large circle (or 'fairy ring') from the active margin of an extensive, perennial, underground mycelium, while the centre of the thallus may be dead or completely rotted away. Parts of a long-lived thallus constantly grow, while other parts are dying; there would seem to be no limit to the survival of the individual thallus under ideal conditions. No doubt nutrition and environment play a great part in the speed with which fruitbodies of fungi are formed. In contrast to soil-inhabiting field mushrooms, the cultivated mushrooms of commerce grow from a well nourished mycelium in an optimum environment; the fruitbodies are formed independently of season from virtually any part of the mycelium near the surface of the mushroom bed, and successive flushes of mushrooms may appear and grow to maturity at intervals of about 10 days to 2 weeks for a period of about 3 months.

In still liquid a mycelium can usually float and form a centrifugally growing mat or pellicle at the surface; but if it is kept immersed by continuously shaking the culture the mycelium tends to grow equally in all directions and to form a roughly spherical thallus; alternatively, since the mycelium is easily fragmented and the pieces can continue growing, a number of smaller thalli may be formed. Shaking not only has the advantage of aerating the mycelium (most fungi are aerobic), but also results in more uniform production of various metabolites which may be required for harvesting and investigation. In still liquid media, and in semi-fluid media such as agar, the immersed and aerial hyphae may often differ in appearance.

Some moulds depend upon light stimulation for sporulation; thus cultures exposed to alternating periods of light and dark may show concentric zones of sporing and non-sporing mycelium which may differ in colour. Diurnal differences in temperature may affect the growth rate and result in alternating zones of luxuriant and sparse mycelium. In species which are not much affected by changes in light or temperature, alternating dense and sparse zones of growth may sometimes be attributed to the effects of staling. Waste products of hyphal metabolism evoke a negatively chemotropic response in hyphae, as shown by their growth away from staled areas. This effect is partly responsible for the centrifugal growth of the thallus, but in addition waste products are accumulated in zones of vigorous hyphal growth and form a chemical barrier in which the growth rate tends to be depressed. The comparatively few hyphal tips which succeed in passing through the barrier eventually reach uncontaminated nutrient sources and again grow and branch profusely. Staling, however, is apparently of little consequence in many fungi; hyphae of the same thallus or from thalli of the same species usually grow close together and, far from showing mutual aversion, may stimulate one another to form anastomosing branches — presumably as a positively chemotropic response. There must be a balance between the attraction to nutrients and to hyphae of the same kind, and the aversion to waste products. Garrett (1963) has summed up these tendencies by stating: 'The net result is that, when a fungal mycelium is growing in an aqueous medium permitting free diffusion of nutrients and growth products, every fungal hypha tends to keep as far away as possible from every other hypha.' In this regard, it is interesting to observe that mycelia of aquatic fungi do not become anastomosed or aggregated into large compact fruitbodies, but instead develop small discrete sporophores; marine Ascomycotina (Johnson and Sparrow, 1961) form fruitbodies from mycelia embedded in wood, or in the tissues of algae, but are unlikely to do so from a free-floating mycelium.

Branching of hyphae

Branching (Fig. 3) is **dichotomous** when the apex of a hypha ceases elongating and forks into two equal branches. More often it is subapical and lateral, leaving the leading hyphal apex free to continue its growth. Lateral branches are usually formed singly, but may sometimes be paired or **opposite**, or may arise in whorls of three or more in which case they are termed **verticillate**. In **sympodial** branching each successive leading apex

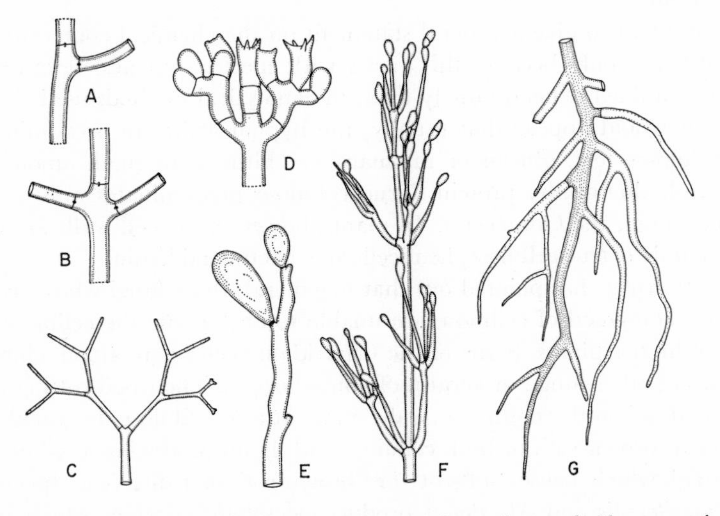

Fig. 3. Branching of hyphae and simple sporophores (*camera lucida*, at varying magnifications). A, simple lateral branching (*Thanatephorus cucumeris*); B, opposite branching (*Thanatephorus cucumeris*); C, dichotomous (*Piptocephalis* sp.); D, cymose (*Botryohypochnus isabellinus*); E, sympodial (*Ramularia* sp.); F, verticillate (*Cladobotryum* sp.); G, monopodial (*Mycotypha microspora*).

becomes restricted in growth and is overtaken by a lateral branch from below; thus a series of superposed branches is formed and may simulate a simple axis. **Monopodial** branching is much commoner than sympodial; here the apex of the leading hypha is not suppressed but keeps pace in growth with the most active of the lateral branches from below, and the impression is given of a single and continuous 'main stem'.

Especially in relation to hyphae forming sporogenous cells the type of branching may be: (*a*) dichotomous; (*b*) verticillate; (*c*) **cymose** — determinate centrifugal branching, i.e. with all axes terminating in sporogenous cells or spores and with the oldest of these at the centre; (*d*) **racemose** — indeterminate and centripetal branching, i.e. with the apex of the main axis sterile and the laterals terminated by sporogenous cells or spores, the oldest of these being towards the base or outside. It will be observed that the cymose fructification results from sympodial branching, while the racemose one is formed from monopodial branching of the sporogenous hyphae.

The hyphal wall

Electron microscopy has shown that the walls of hyphae are laminated, usually being composed of two, or sometimes several layers of microfibrils arranged in various ways in an amorphous matrix. This type of construction is also found in the cell walls of plants, but with differences in chemical composition.

It is difficult to give a general statement on the chemical composition of cell walls, not only because this varies with species, but also because the answers obtained depend largely upon the analytical methods used. Nevertheless, it would appear that usually, the hyphal walls are predominantly hemicelluloses (e.g. glucan or mannan) or chitin, with small amounts of lipids and, doubtfully, protein. True cellulose predominates only in one group of fungi, the Oomycetes. In plants, however, the cell walls are composed mainly of true cellulose, hemicellulose, pectin and lignin.

Garrett (1963) has pointed out that in general, those fungi whose hyphal walls are composed of cellulose are unable to decompose the cellulose and lignin of higher plants. Some of the Chytridiomycetes may attack chitin in the walls of other fungi, or keratin of animal origin. The wood-rotting fungi with predominantly chitinous walls may attack cellulose in wood and produce a 'brown-rot', or both cellulose and lignin producing a 'white-rot'. The fungi which cause 'soft-rot' or 'brown-rot' of fruits (e.g. species of *Rhizopus*, *Botrytis* and *Monilinia*) produce pectolytic enzymes which break down the pectates in the middle lamellae and walls of host cells prior to lysis of the cell contents. The innermost layer of the hyphal wall adjoins the **plasmalemma**, the membrane which surrounds the cytoplast.

The walls of fungal hyphae may be uncoloured (**hyaline**), or darkened with yellowish or brownish pigment allied to melanin. Reddish, orange, greenish, bluish or purplish pigments occur in some fungi, but not usually as part of the wall substance. In some instances pigments which colour the whole mycelium are found dissolved in the cytoplasm. Some fungi contain oxidase enzymes which change the colour of the flesh when it is cut and exposed to air, or bruised; many of the boleti react in this way. Some mushrooms, notably species of *Lactarius*, contain laticiferous hyphae which exude drops of latex when they are damaged. This may be watery or milky, white or coloured, and its colour may change on exposure to air. Somewhat similar hyphae, but with tannin-like contents which 'bleed' bright red in contact with air, are to be found in polypores such as *Amauroderma rude*, and in *Stereum sanguinolentum*. Several of the higher fungi contain luciferin and luciferase and become strongly luminescent under suitable conditions of growth; the glow emitted by a cluster of fruitbodies of the mushroom *Pleurotus nidiformis* is sufficient to read by, or to allow a photograph of the fungus to be taken by its own light. Crystals, often of calcium oxalate, are not uncommon among Basidiomycotina, forming an encrustation on hyphae and on certain other sterile structures in the fruitbodies.

Septation of hyphae

Transverse walls or **septa** (Fig. 4) occur in the hyphae of all filamentous fungi, dividing them into a number of interseptal compartments known as cells, which are joined end-to-end in a linear series. The septum is formed by an annulus of wall material which invaginates the plasmalemma and grows centripetally inwards; its manner of growth is analogous to the closing of an iris diaphragm.

Much remains to be discovered about septation, but from the information at present available it would seem that there are two general types of septa: primary and adventitious. They are defined cytologically and the reader may find it helpful to refer to pp. 75–77 for an explanation of some of the terms now to be used. **Primary septa** are formed in association with true mitotic or meiotic nuclear division, and they separate the daughter nuclei. **Adventitious septa** are formed in the absence of true mitosis or meiosis and occur especially in association with change in the local concentration of cytoplasm as it moves from one part of the fungus to another (Talbot, 1968). The difference is well shown in *Basidiobolus ranarum* (Entomophthorales; Robinow, 1963). In this species the cytoplasm remains concentrated towards the apex of a growing hypha so that as the apex advances, the proximal part of the hypha becomes almost devoid of cytoplasm. Then, a transverse septum is formed along the lower edge of the contracted protoplast and separates the empty part of the hypha from its distal part occupied by the mass of cytoplasm. 'Meanwhile at the far side of the septum, the cytoplasm moves forward again. . . . A single moving protoplast thus leaves in its wake a long chain of empty cell chambers' (Robinow, 1963). The original cell has a single large nucleus which moves forward with the cytoplast; thus the septa just described are formed without association with nuclear division, and may be called adventitious. The protoplast itself gradually increases in size and eventually a state is reached where the nucleus divides by true mitosis accompanied by the laying down of a transverse septum which segregates the two daughter nuclei into two viable cells full of cytoplasm. The lower of these often branches, and both cells then continue to advance and lay down further adventitious septa. It is clear, however, that the second type of septum, formed in association with mitosis of the nucleus, is cytologically different from the adventitious septa formed when cytoplasm is withdrawn: it is instead a primary septum.

In some other Entomophthorales, Mucorales and Saprolegniales (Robinow,

1957*a*, 1957*b*, 1963) the individual somatic cells are multinucleate and their nuclei divide by **constriction**, not by a true mitosis involving the formation of a spindle and a metaphase plate. Therefore, any septa that are formed cannot be associated with mitosis; instead they appear to be initiated in any

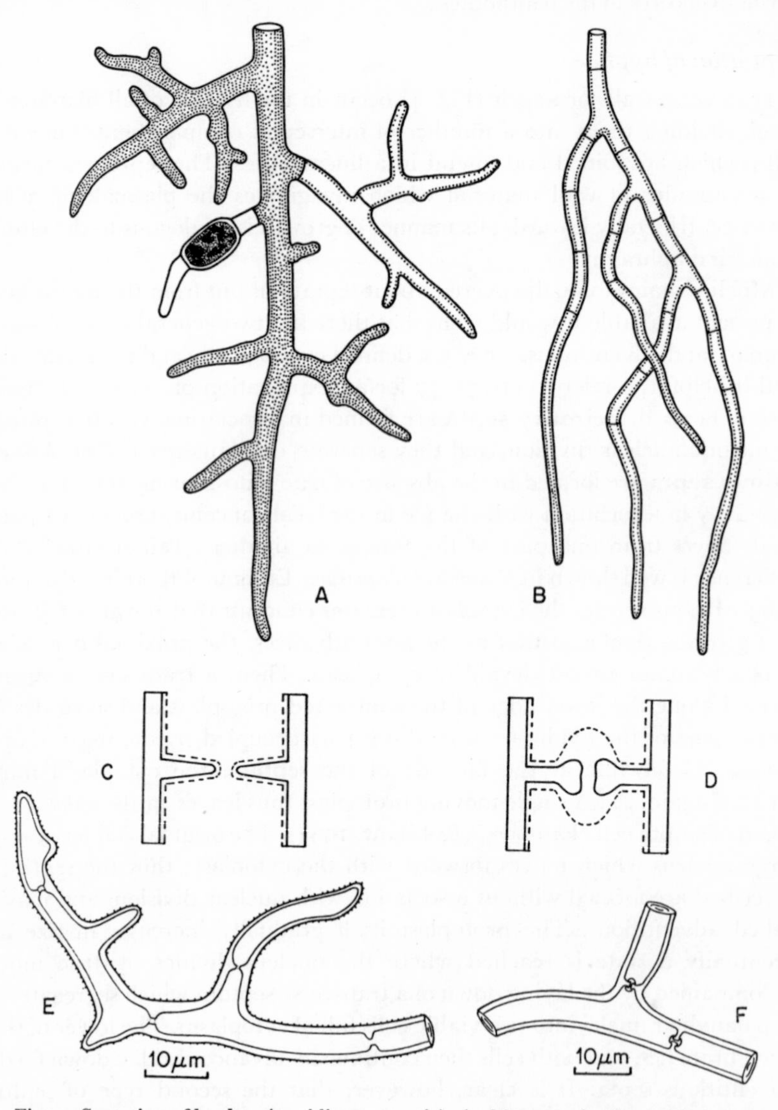

Fig. 4. Septation of hyphae (semidiagrammatic). A, *Mucor* sp., hyphae with sparse adventitious septa, and a chlamydospore; B, *Phialomyces macrosporus*, hyphae with regular primary septa; C, septal pore of Ascomycotina; D, dolipore septum with septal pore cap in Basidiomycotina; E, *Gymnoascus reesii*, hyphae with open septal pores; F, *Ceratobasidium cornigerum*, hyphae with doli ore septa.

part of the mycelium where the local concentration of the cytoplasm is greatly altered by accumulation or withdrawal. Such septa may be formed when cytoplasm regresses from branch hyphae no longer in contact with food supplies, when cytoplasm is withdrawn from old or wounded parts of the mycelium, or when it is withdrawn from a main axis and accumulates in reproductive cells. In general such septa are irregular and sparse, but they may sometimes be formed in a series as the cytoplasm moves forward. These septa are evidently adventitious. The cells from which cytoplasm has regressed do not necessarily collapse; they may remain turgid, with a very thin peripheral layer of cytoplasm and a vacuole filling the rest of the cell. They may still have important functions, e.g. the inflated cells which compose the flesh of some fruitbodies.

In hyphae of higher fungi with uninucleate or dicaryotic cells, the septa present are associated with true mitosis, and this implies that they should usually be formed close behind the growing apex, appearing regularly and in acropetal succession and producing a septate hypha with cells of roughly comparable size. They are primary septa. We have little information on nuclear division and septum-formation in hyphae of higher fungi with multinucleate cells. In *Thanatephorus cucumeris* (Flentje, Stretton and Hawn, 1963) the young apical cell of a hypha commonly has five–eight nuclei, while old cells further back in the hypha have two–four. Two types of septa are formed, though not necessarily always in exactly the same way: primary ones associated with simultaneous mitosis of the nuclei in a growing apex, and subsequent separation of the daughter nuclei; second a thinner type of septum, presumably adventitious, which later divides old cells and separates their sister nuclei without mitosis. In the fruitbodies of some higher fungi, Corner (1950) has noted that inflated cells become divided by thin membranous secondary septa.

Primary septa predominate in higher fungi (Ascomycotina, Basidiomycotina and their asexual states). In these fungi there is an open **septal pore** (sometimes more than one) at the centre of the septum, and the plasmalemma is continuous around the lip of the pore (Fig. 4). Fluid and some of the smaller cell organelles are able to pass between adjacent cells; even nuclei are sometimes able to migrate from cell to cell, although this is probably uncommon. In Ascomycotina the primary septa are of uniform thickness or are tapered towards the central pore; there is usually a single simple pore. However, some species are known in which the septum has several perforations around its periphery as well as a few nearer the centre, each with one or more adjacent dense 'Woronin bodies'. In a species of *Fusarium*, Reichle and Alexander (1965) obtained evidence that the Woronin bodies may plug the perforations and probably seal off degenerating or damaged cells. In all Basidiomycotina that have been investigated, except in the rusts (Uredinales), the basic design of the septal pore is greatly modified by associated structures, and a **dolipore septum** is the result (Fig. 4). This is a primary septum, greatly expanded into a thickened ring surrounding the central pore; in section the ring appears like a thick pad on either side of the septal pore. On both sides of the septum, covering the pore and ring, is an arched **septal pore cap** which in some cases (Wilsenach and Kessel, 1965) is perforated with as many as twenty — fifty small holes so that it has a sieve-like appearance. Adventitious septa are comparatively rare in higher fungi, but have been noticed in connexion with change in density of cytoplasm in hyphae (especially skeletal hyphae), cystidia, basidia and sterigmata.

In the lower fungal groups (Mastigomycotina and Zygomycotina) primary septation must be rare, at least in those fungi whose nuclei divide by constriction, and in which adventitious septation is the general rule. The fine structure of adventitious septa does not appear to have been determined yet, either in lower or higher fungi. We have assumed above that all adventitious septa are homologous, but it is of considerable taxonomic importance to know whether this is true or not.

THE FUNGAL CELL

Although minor differences in composition and arrangement do occur, the fine structure (Hawker, 1965; Aronson, 1966; Bracker, 1967) of fungal cells (Fig. 5) is broadly the same in both unicellular and filamentous thalli of those fungi that have been investigated. The most notable exceptions occur in some of the Oomycetes.

The structure of the cell wall has already been described. Adjacent to this is the plasmalemma, a membrane forming part of the cytoplasmic membrane system and enclosing the whole cytoplast. Other cellular inclusions, bounded by or associated with membranes, are the mitochondria, dictyosomes and their tubular or vesicular derivatives, vacuoles, ribosomes and nuclei. The non-membranous inclusions within the cell include granules, lipid bodies and the endoplasmic matrix. Each vacuole is surrounded by a membrane, or tonoplast. Many mitochondria are usually present in the cell, each with a three-layered membrane whose innermost layer becomes infolded to form finger-like or plate-like series of cristae mitochondriae. An endoplasmic reticulum is present only in irregular segments. The cell contains one or more nuclei each consisting of nucleoplasm surrounded by a porous nuclear membrane; sometimes the nucleoplasm shows a dense area considered to be the nucleolus, and irregular patches which are possibly chromatin. An extra, parallel, nuclear membrane sometimes covers the area occupied by a pore in the main nuclear membrane. Lomasomes, or small pockets adjacent to the wall, are formed by invagination of the plasmalemma. In some instances plasmodesmata have been observed in the septa of fungi.

In their fine structure, the cells of fungi and plants are very similar, but

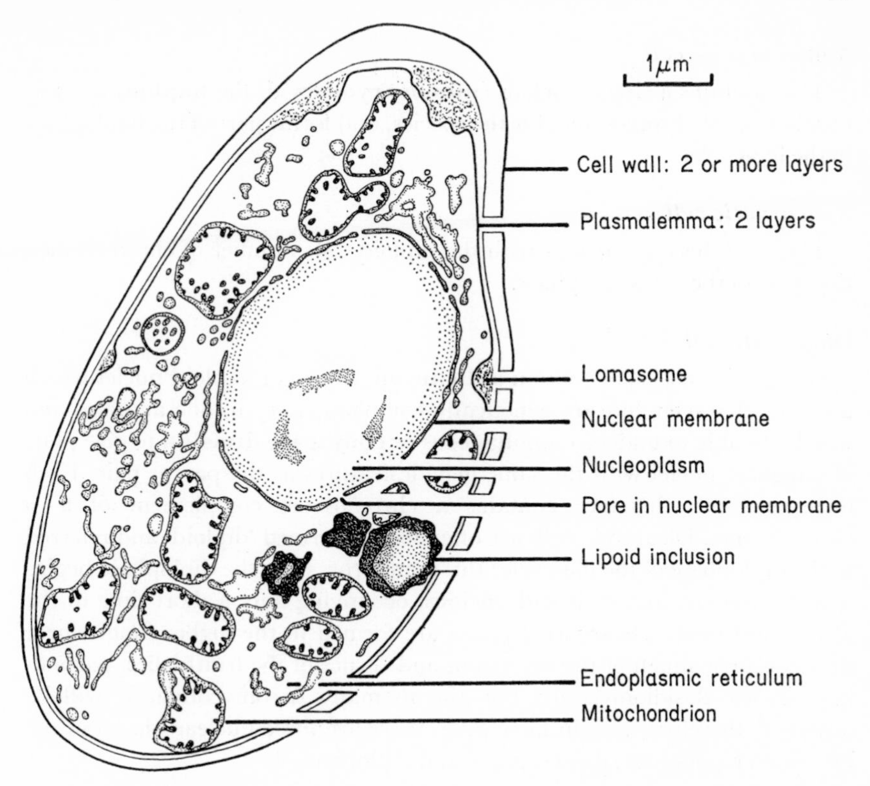

Fig. 5. A fungus cell, drawn from an electron micrograph of an haustorium of *Peronospora parasitica*. (Micrograph by courtesy of Dr Elaine Davison.)

very different from those of bacteria and Actinomycetes in which no mitochondria are found and the nuclear material is scattered, not enveloped by a nuclear membrane. There are, however, some minor differences between cell structure in fungi and plants. The endoplasmic reticulum is irregular in fungi, and ribosomes are not attached to the reticulum but occur throughout the endoplasmic matrix (Hawker, 1965). Nuclear division is simpler in some fungi than in others; in some it occurs by constriction rather than true mitosis. Finally, as has been mentioned before, there are differences in cell wall composition in fungi and plants.

NUCLEAR COMPLEMENT AND NUCLEAR BEHAVIOUR OF CELLS

Certain terms are used to describe the nuclear complement and behaviour of fungus cells.

Monocaryotic cell

This contains a single nucleus (**monocaryon**) with the **haploid** or basic number (n) of chromosomes for the species, and forms part of the haplophase in the life-cycle.

Syncaryotic cell

This contains one nucleus with the diploid ($2n$) number of chromosomes, thus part of the true diplophase.

Dicaryotic cell

Dicaryotic cells each contain a **dicaryon**, a pair of haploid nuclei which differ in character but are genetically complementary to one another, and which are able to undergo simultaneous or **conjugate division** to give pairs of daughter nuclei with the same genetic features as the parent pair. Each dicaryotic cell therefore contains a chromosome complement of $n+n$ chromosomes. Dicaryotic cells are often loosely termed 'diploid' and referred to the diplophase of the life-cycle; this is an error, since the diplophase begins with karyogamy and ends with meiosis, occupying only a short time in the life of most fungi. Dicaryotic hyphae are formed in the higher fungi where they comprise much of the mycelium and tissues of the fruitbodies, particularly in the Basidiomycotina but also in many Ascomycotina. It will be observed, therefore, that in some fungi there are three nuclear phases in the life-cycle: haplophase, dicaryophase and diplophase.

Dicaryotization

This is the process by which a haplophase cell or mycelium becomes dicaryotic with a pair of conjugate haploid nuclei in each cell. This process is often incorrectly called diploidization. It will be discussed in some detail later.

Heterocaryotic and homocaryotic cells

A heterocaryotic cell contains at least two genetically different types of nuclei (**heterocaryons**). The term heterocaryotic is often applied to hyphae with multinucleate cells, but these are not necessarily heterocaryotic since the nuclei may all be of one genetic type, i.e. **homocaryotic**. It should also be noted that the dicaryotic condition is a special case of the heterocaryotic, and here the cells are only binucleate.

The cells of heterocaryotic hyphae may differ in the number, kinds, and proportions of each type of nucleus present; even mixtures of haploid and diploid nuclei may be encountered. It will be apparent that the great variations possible in nuclear complement can bring about great morphological and physiological variation between different thalli of the same species, when such thalli have originated from different spores into which unlike nuclei have been segregated during spore-formation. Heterocaryons usually arise by anastomosis of hyphae or from mutations, and may also change their nuclear complements by these processes; cells or tissues growing from anastomosed or mutated cells may produce **sectors** of the thallus with changed physiological or morphological properties. Genetic variation of this sort is usually permanent and inheritable. Homocaryotic hyphae, whose cells may be multinucleate but contain only one genetic type of nucleus, give rise to spores or other cells which are less variable as a result of genetic causes; but the resulting thalli may nevertheless show equally great temporary variation arising from the effects of environmental factors.

SPECIALIZED SOMATIC STRUCTURES

In fungi with a mycelial thallus, all specialized structures, whether assimilative or reproductive, are modifications of hyphae or groups of hyphae. This suggests that we need a good knowledge of hyphae and their general properties but regrettably, far less attention is ever paid to the humble hypha than to the much more spectacular reproductive structures. The reason for this is not hard to find: in general hyphae are diffuse and difficult to investigate, and their taxonomic features are hard to pinpoint and describe. A certain amount — but not enough — is known about the potentialities of growth-forms constructed of tubular elements, and about the effects of various ecological situations in which they occur in shaping the morphology and function of fungal structures. Since no fungal structures should be considered in isolation, or from a purely morphological point of view, an attempt is made in a later chapter (Chapter 13) to correlate some of these factors. Meanwhile the approach has to be largely one of descriptive morphology.

Rhizoids

A rhizoid (s; Fig. 6) is a short, root-like filamentous branch of the thallus, generally formed in tufts at the base of small unicellular thalli or small sporophores, serving as a holdfast and also capable of absorbing water and nutrients. Rhizoids are often connected by aerial **stolons**, or runner-hyphae, to other groups of rhizoids, e.g. in *Rhizopus*. They appear to be associated only with some of the lower fungi (Mastigomycotina and Zygomycotina) and may occur in either saprobic or parasitic species. In some of the monocentric Chytridiomycetes there may be extensive rhizoidal systems, e.g. in *Rhizophydium*. In some of the polycentric Chytridiomycetes the rhizoidal system is still more extensive and resembles a mycelium; narrow, non-nucleated rhizoidal strands become interconnected and in parts develop nucleated expanded portions which may produce further rhizoids or become differentiated into sporangia or resting-spores. This type of polycentric rhizoidal thallus is known as a **rhizomycelium** (Karling, 1932).

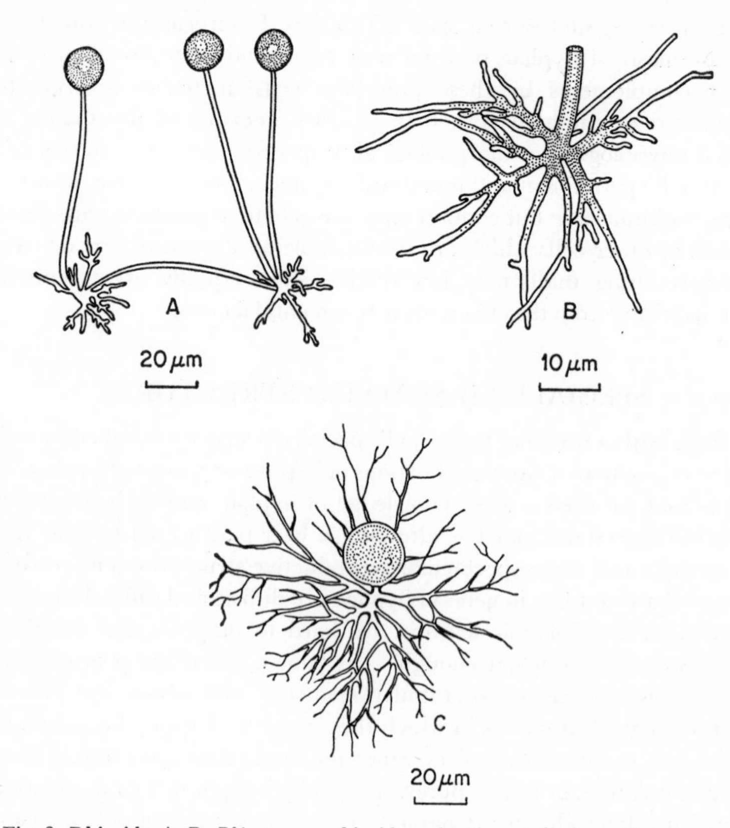

Fig. 6. Rhizoids. A, B, *Rhizopus* sp., rhizoids and stolon at the base of sporangiophores; C, *Rhizophydium sphaerotheca*, rhizoids at the base of a sporangium.

The following three types of specialized structures are associated with some parasitic fungi. They are concerned with attaching the fungus to the host surface, penetrating host tissues, or extracting nutrient from the host tissues.

Appressoria

An appressorium (s; Fig. 7) is a simple or lobed mucilaginous swelling on a germtube or hypha, attaching these to the surface of the host or other substratum. Appressoria are formed by some types of parasitic fungi (e.g. Erysiphales) usually at an early stage of infection of the host, but are also formed by some other types of fungi whose germtubes or hyphae are in contact with a hard surface. Besides attaching the hypha, the appressorium of a parasitic fungus assists the fine infection pegs, or penetrative branches, to pierce the host cuticle by providing a firm hold which counteracts the force of penetration.

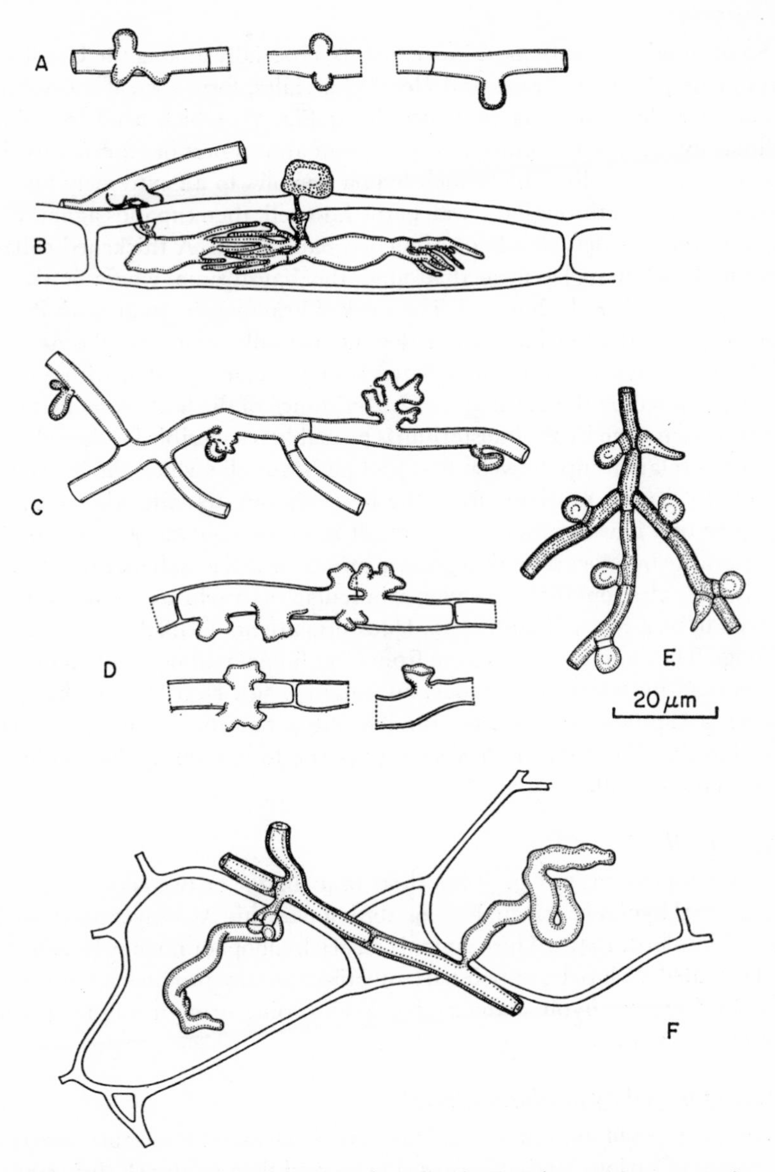

Fig. 7. Appressoria, haustoria and hyphopodia. A, *Erysiphe graminis*, appressoria; B, *Erysiphe graminis*, appressoria, penetration pegs surrounded by callose sheath, and large digitate haustoria; C, *Microsphaera alphitoides*, appressoria and small globose to ellipsoid haustoria; D, *Microsphaera polonica*, appressoria; E, *Meliola* sp., hyphopodia; F, *Uromycladium tepperianum*, intercellular hypha and haustoria.

Haustoria

Some fungi which are parasitic on plants, especially among the Uredinales, Erysiphales, Peronosporales and Hemisphaeriales, form special intracellular branches called haustoria (s. haustorium; Fig. 7), which arise from intercellular hyphae, or from infection pegs from appressoria or external hyphae, and invade host cells. The branch hypha narrows to an extremely fine diameter as it passes through the wall of the host cell, then expands again within the cell into a wider, simple or branched haustorium. A thickened collar of host material may be formed around the haustorium where it emerges through the wall of the host cell. The work of Fraymouth (1956) with Peronosporales shows that the haustorium does not actually pierce the plasmalemma of the host cell, but simply invaginates it and becomes coated with a callose sheath secreted by the plasmalemma. Resistance of the host plasmalemma to actual penetration by the haustorium is probably responsible for the delicately balanced relationship between host and pathogen, by which the haustorium is able to extract nutrients from the host without actually killing it. The presence of a haustorium does not result in better contact between the host and parasite but does provide a greater surface area for exchange of materials. The arbuscules found in vesicular-arbuscular mycorrhizas (p. 68) are considered to be a type of haustorium. Haustoria are not formed in agar cultures of fungi. They appear to be absent from most fungal pathogens of animals but are present in some of the predacious fungi (p. 59). Here, the hyphae penetrating a captured animal may branch into a haustorium-like structure or may form an 'infection bulb' which gives rise to branch hyphae within the body of the animal.

Hyphopodia

A hyphopodium (s; Fig. 7) is a short branch, one or two cells in length, of an external hypha in certain leaf-inhabiting parasitic Ascomycotina (Doidge, 1942; Hansford, 1946). The terminal cell of a hyphopodium may be expanded and rounded, or lobate, or pointed; sometimes it may produce a haustorium. Hansford regards hyphopodia as special absorbing structures in fungi whose mycelia are mainly external to the host.

Snares formed by predatory fungi

Several types of soil-inhabiting fungi have been found to capture nematodes by means of various kinds of snares (Fig. 8) and then to absorb their contents through branches which ramify through the body of the dead animal. The simplest form of snare is a sticky hypha, or a small sticky bulb formed as a lateral branch of a hypha. More complex is the type of snare formed by hyphae which branch, coil and anastomose to form a series of interlaced loops or simple rings which capture the nematodes by adhesion. The most spec-

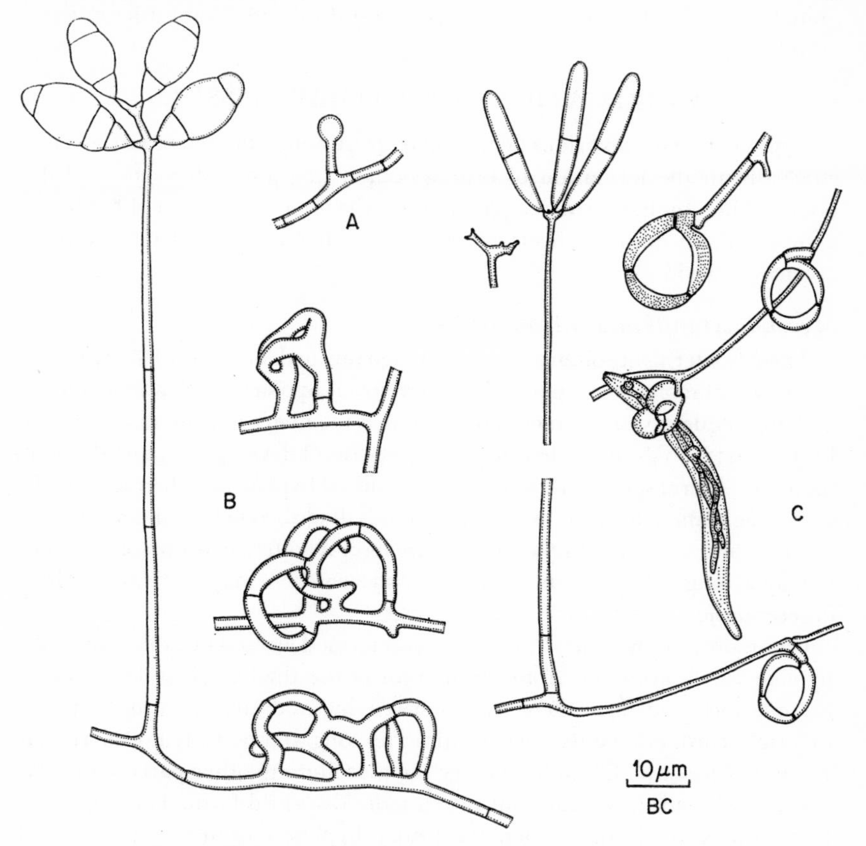

Fig. 8. Snares of predatory fungi. A, small sticky bulb (diagrammatic); B, interlaced anastomosing loops (*Dactylaria thaumasia*); C, inflating and constricting rings (*Arthrobotrys dactyloides*).

tacular snares, however, are specialized rings formed on short lateral branches and composed of three cells which are capable of instant inflation on being touched on their inner surfaces. A nematode poking its body through one of these rings is immediately constricted and held tightly by the inflated sensitive cells. Usually, these snares are formed in response to the presence of nematodes, but they can also be induced to form in their absence by including various chemicals or animal proteins in the culture medium.

Most of the fungi which capture nematodes are members of the Hyphomycetes (Deuteromycotina). Some of the Zoopagales (Zygomycotina) are able to prey on nematodes but are much commoner as predators of soil amoebae, rhizopods or rotifers. These Zoopagales apparently do not have special snares, but their conidia may be ingested and germinate inside the animal, or may stick to its body and penetrate it after the formation of a germtube.

AGGREGATIONS OF HYPHAE: TISSUES

Hyphae may show various degrees of aggregation, adhesion or coalescence, especially in the formation of resting-bodies, migratory strands and fruit-bodies. The simplest form of aggregation is the loose association of hyphae as a mycelium. Compact sheets of mycelium often form under bark or in crevices in wood.

Mycelial strands and rhizomorphs

These two terms are often used interchangeably to denote cord-like structures composed of more or less parallel or interwoven hyphae which adhere closely and are frequently anastomosed or cemented together. Townsend (1954) and Butler (1957, 1958) have described the patterns of development and differentiation in several species. In some, the individual hyphae lose their identity by becoming compacted into tissues of the kinds described below under 'plectenchyma', and a hard, compact cortex encloses a softer, whitish core of more loosely arranged hyphae; but with other species there may be relatively little differentiation into distinct zones or tissues.

Commonly, a mycelial strand (Fig. 9) is formed around one or more leader hyphae which grow out from the margin of the thallus; these leaders then become surrounded by their own interweaving and anastomosing branches to form a cord, often only a few centimetres long and possibly not more than 1–2 mm thick, which is able to carry the mycelium to other parts of the substratum where it again fans out into a more dispersed form. The suggestion that strands are formed when the leader hyphae encounter surfaces with

Fig. 9. Hyphal strands of *Peniophora filamentosa* (diagrammatic).

little free nutrient and fan out again on more nutritious parts of the substratum, may not be entirely correct, since although strands are uncommon in fungi grown on nutrient media they undoubtedly do occur in some species in culture (e.g. in *Sclerotium rolfsii*). Again, the suggestion that strands may be formed in response to limitation of water and that even a narrow strand would probably resist desiccation better than dispersed hyphae is contradicted by the behaviour of some species in culture, where water is usually not a limiting factor. Evidently, more work needs to be done on the factors which contribute to strand-formation. Garrett (1963) points out that most strand-forming fungi are cellulose-decomposers. He suggests that strand-formation is an advantage where there is competition between cellulolytic fungi in colonizing a woody substratum, since an inoculum consisting of a strand can be expected to contain a greater bulk of hyphae and greater potential energy (related in this case possibly to cellulase content) than one in the form of dispersed hyphae.

A different type of development is found in the cords of *Armillariella mellea*, and Garrett would limit the term 'rhizomorph' to cords of this sort. Parallel, unbranched hyphae become closely associated and grow as a single unit from an apical meristematic region. In this co-ordinated manner of growth and in the production of endogenous branches of similar construction, these rhizomorphs are root-like. Garrett observes that they differ from mycelial strands not only in construction, but also in being produced freely on nutrient media and not fanning out into discrete hyphae. Presumably because their hyphae are unbranched such rhizomorphs grow much faster than unorganized hyphae of the same species. In *A. mellea* the rhizomorphs may be up to 5 mm wide, and Findlay (1951) has recorded an instance where they reached a length of 9 m growing in a water tunnel. They are blackish and hard on the outside and resemble a bootlace.

Rhizomorphic structures are apparently limited to the higher fungi, and particularly to the Basidiomycotina. It is interesting to note, in passing, that the nests of some sun-birds in South Africa are made almost entirely of the wiry black rhizomorphs of a species of *Marasmius*, a small mushroom.

Plectenchyma tissue

Plectenchyma (Fig. 10) is the name given to any tissue organized from fungal hyphae. When the tissue is rather loosely woven of easily distinguishable hyphae it is called **prosenchyma**. When the hyphae become short-celled and so closely aggregated that the cells are rounded or angular by mutual pressure, the tissue is known as **pseudoparenchyma** if the cells are thin-walled, or **pseudosclerenchyma** if they are thick-walled and dark in colour. Finer distinctions are sometimes made in describing the composition of fruitbodies of the Ascomycotina and Basidiomycotina, in which tissue-formation reaches its highest level.

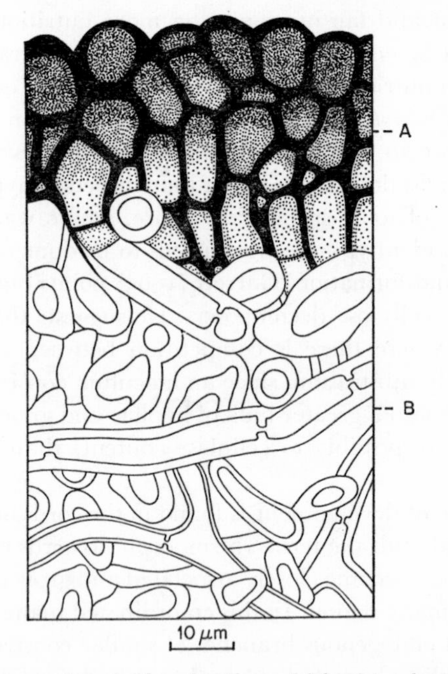

Fig. 10. Plectenchyma tissue in the sclerotium of *Sclerotinia sclerotiorum*. A, pseudosclerenchyma; B, prosenchyma, with easily visible septal pores.

Sclerotia

Sclerotia (*s.* **sclerotium**; Fig. 11) are resting-bodies formed by aggregation of somatic hyphae into dense rounded, or sometimes flattened or elongated, masses. Sometimes they are more or less characteristic for a particular species, in size, shape and colour, within a wide range of variation. The small sclerotia (microsclerotia) of *Macrophomina phaseoli* may be less than 100 μm in diameter, and the majority of sclerotia of other species do not exceed a maximum dimension of about 2 cm, although some species of polypores produce sclerotia which weigh several kilograms and may be 25 cm or more in diameter. Such sclerotia formed by *Polyporus mylittae* used to form part of the diet of Australian aborigines. Sclerotia serve as storage and survival structures which can tide the fungus over periods of drought, cold or moderate heat; they also serve as propagules from which new mycelia can grow. Most soil fumigants are capable of killing all but a small proportion of sclerotia in soil, but the sclerotia of some important pathogenic species of *Verticillium* are known to survive methyl bromide treatment.

Sclerotia of many species are formed in well nourished active cultures, which suggests that their formation is not related to food shortage. Townsend and Willetts (1954) traced the initiation of sclerotia in several species and found that they could develop in three different ways. In the 'loose' type of

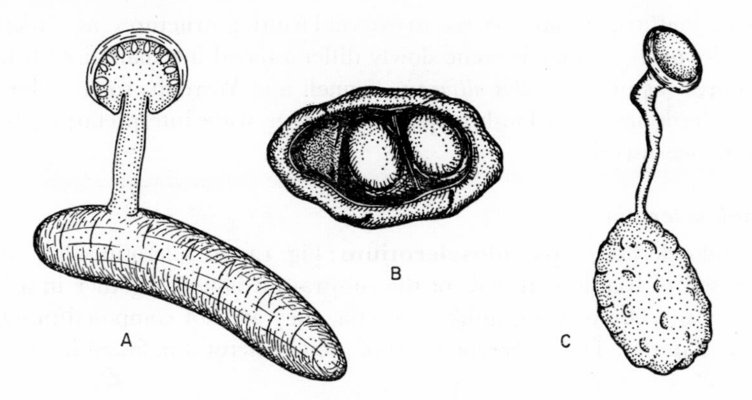

Fig. 11. Germination of sclerotia. A, *Claviceps purpurea*, stipitate stromatic ascocarp formed externally; B, *Aspergillus alliaceus*, with internally-formed cleistothecia; C, *Sclerotinia sclerotiorum*, with externally-formed apothecium. (B, after photograph by Fennell and Warcup in *Mycologia* 51, 410, 1959.)

development, exemplified by *Rhizoctonia solani*, the sclerotial initials are formed from adjacent hyphae which produce moniliform branches, i.e. branched chains of short barrel-shaped cells formed in acropetal sequence. These hyphae aggregate but do not coalesce except towards the centre of the sclerotium, which thus remains relatively soft and homogeneous and retains its hyphal characteristics — at least towards the outside. Moniliform hyphae of this type are found in the rhizoctonia-like imperfect states of some species of the genera *Sebacina*, *Tulasnella*, *Thanatephorus*, *Waitea* and *Ceratobasidium* and sometimes, but not always, continue their development to form sclerotia (Warcup and Talbot, 1962, 1966, 1967). In the 'terminal' type of sclerotium-formation, the apex of a single hypha branches dichotomously and septa cut the branches off as two short cells which repeat the dichotomy; this process continues at each new branch apex thus formed and the whole structure eventually becomes compact and coalescent to form the sclerotium, which, when mature, is differentiated into an outer rind, a narrow cortex of pseudo-parenchyma cells and a large central medulla of prosenchyma cells. The central mass ceases growth when the rind has been differentiated. In the 'strand' type of development, intercalary parts of one or more adjacent hyphae in a strand produce numerous lateral branches which septate into short cells and also continue branching and intertwining. The central mass of cells coalesces, but growth continues from this part and stretches the outer walls of the sclerotium. The sclerotium is zoned, as in the 'terminal' type, but differs in that the rind is narrow and the cortex broad. In these zonate types of sclerotia the rind and the outer cortical cells are usually pigmented, forming a protective layer outside the central storage cells.

On 'germinating', sclerotia may produce a new mycelium directly (as in

Sclerotium rolfsii), or may give rise to external fruiting structures (as in *Claviceps* and *Sclerotinia*), or may become slowly differentiated internally into fruiting structures (as in *Aspergillus alliaceus*; Fennell and Warcup, 1959). Thus the term sclerotium is obviously capable of rather wide interpretation, but is nevertheless useful.

Pseudosclerotia

Pseudosclerotia (s. **pseudosclerotium**; Fig. 12) are sclerotium-like masses composed of friable materials of the substratum bound together in a hard mass by mycelium; thus, unlike sclerotia, they are not composed purely of fungal material. The commonest sort of pseudosclerotium, found in saprobic

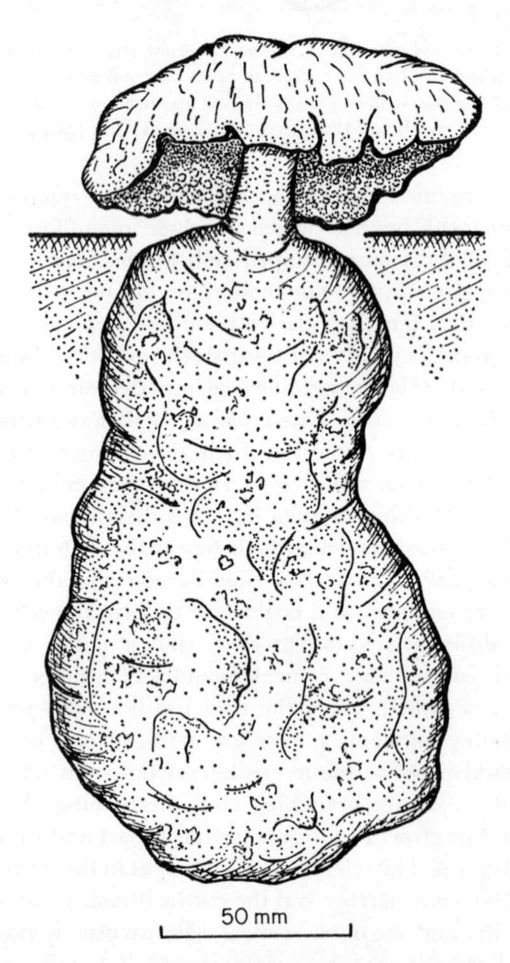

Fig. 12. Pseudosclerotium of *Polyporus basilapiloides* with basidiocarp formed above soil level.

species, is composed of sandy soil bound by mycelium, and it occurs underground at the base of various types of fruitbodies of higher fungi. In the Australian 'stone-making fungus' (*Polyporus basilapiloides*) these structures are sometimes as much as 10 cm in diameter and 20 cm long; they occur in mallee scrub country and, with the first rains after a bushfire, produce fruitbodies above ground (Cleland, 1934). Smaller pseudosclerotia are not uncommon with some species of mushrooms, morels and Gasteromycetes, especially when these are growing in sand. In some parasitic species of fungi, on the other hand, a pseudosclerotium consists of a mycelium and the remnants of plant or animal tissue of the host. The bodies of caterpillars parasitized by *Cordyceps*, and stone fruits parasitized by *Monilinia*, become hardened and 'mummified' by mycelium. Such 'mummies' are pseudosclerotia which eventually may give rise to fructifications.

Pseudorhizas

The mycelium of some mushrooms originates from materials buried a considerable distance underground and, in becoming elevated to the surface to form the fruitbodies or 'mushrooms', it becomes aggregated into one or more stout, subcylindrical columns each of which bears a fruitbody at ground level. Viewed from the position of the fruitbody, the pseudorhiza (s; Fig. 13) is a root-like continuation of the stipe of the mushroom, connecting it with specific underground sources of nutrient. In *Oedemansiella radicata* the pseudorhiza originates from roots of higher plants; in *Collybia fusipes* it comes from roots or buried wood and divides into perennating branches near its apex, from which successive crops of mushrooms are formed (Buller, 1934). The pseudorhizas of species of *Termitomyces* (Fig. 13), which are common in Africa, originate from the comb of living termite nests and sometimes traverse as much as 120 cm before reaching the soil surface; in some species, the 'mushrooms' which they bear are as large as a dinner plate — and often find their way into the dinner!

Stromata

The term stroma (s; Fig. 14) is applied to compact hyphal matrices, which sometimes include fragments of substratum, in or on which fructifications are formed. Details of some types of stromata will be given later. For the moment it may be noted that some sclerotia are essentially stromatic since they form fruitbodies on germination; the other types of sclerotia, which germinate directly into mycelia, can be regarded as 'infertile stromata'. Many rhizomorphs are also stromatic in structure despite the fact that they do not give rise to fructifications. It is therefore important to include the function of producing fructifications in the definition of a true stroma.

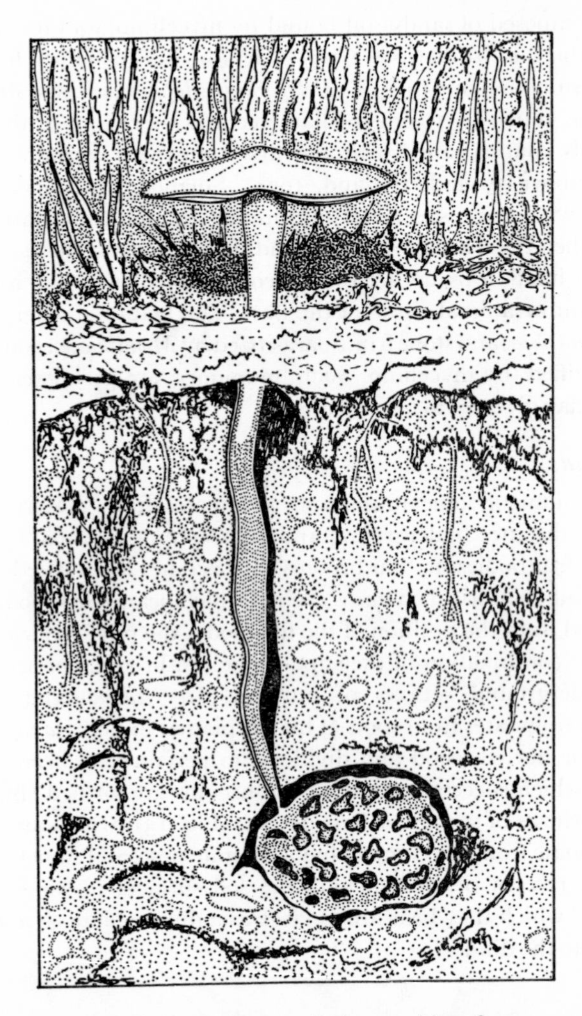

Fig. 13. A mushroom, *Termitomyces cartilagineus*, arising from a pseudorhiza originating in a termite comb about 0.9 m below soil level. (Based on a photograph at the Botanical Research Institute, Pretoria, S. Africa).

MYCORRHIZAS

A mycorrhiza is a compound organ formed by the association of fungal hyphae with a root of a higher plant in a symbiotic, non-pathogenic or weakly pathogenic relationship. The presence of the fungus may sometimes be essential to good growth of the plant, assisting in mineral uptake by its roots, but sometimes it has little or no clear effect. Some species of higher plants, notably trees, form mycorrhizas with several different species of fungi; other

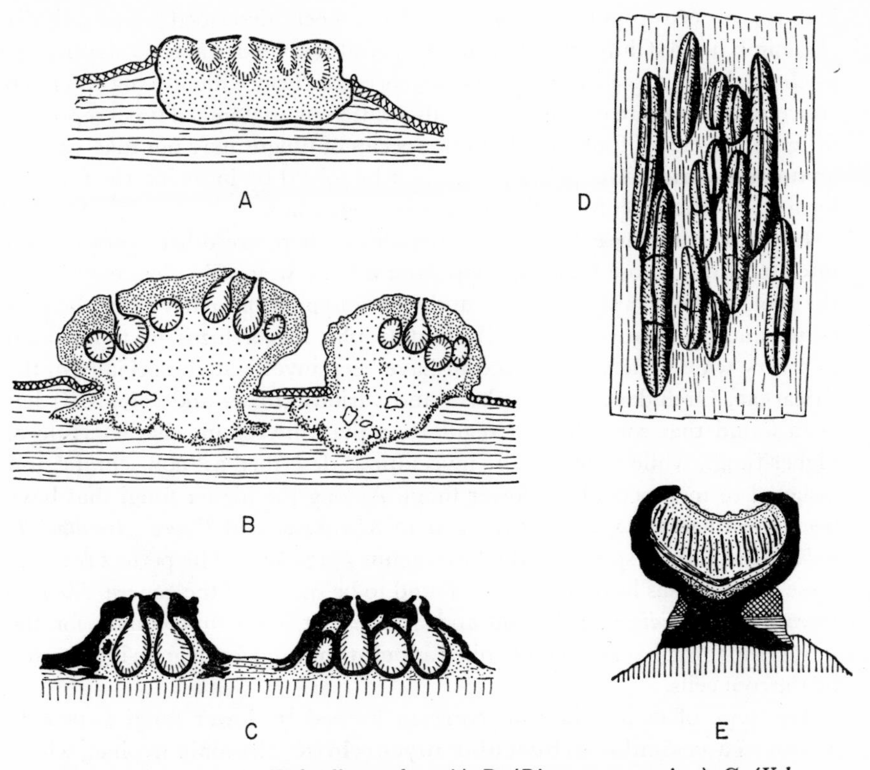

Fig. 14. Stromata. A, (*Holstiella usambarensis*), B, (*Diatrype macowaniana*), C, (*Valsaria eucalypti*), showing perithecia embedded in stroma arising from host (*camera lucida* drawings at various magnifications); D, hysterothecia of *Hysterographium* sp. (diagrammatic); E, hysterothecium in transverse section.

plants are more specific in their associated fungus. Harley (1959) gives a useful summary of our knowledge of mycorrhizas.

The fungus of an **ectotrophic** mycorrhiza forms a sheath of pseudoparenchyma around the root surface, with branch hyphae sometimes forming an intercellular network known as a 'Hartig net' between the outer cortical cells of the root. These mycorrhizas have been investigated mainly in connexion with forest trees, especially *Pinus* and *Fagus*. In the beech, some of the hyphae penetrate host cells giving rise to the **ectendotrophic** type of mycorrhiza. By field observation of species of fungi associated with these trees, and by testing their ability to form mycorrhizas with roots in culture, it has been found that the majority of fungi forming ectotrophic mycorrhizas are Basidiomycotina, with species of *Amanita*, *Tricholoma* and Boletaceae predominating; some Gasteromycetes (especially species of *Scleroderma* and *Rhizopogon*) and a few Ascomycotina are also represented. A member of the Fungi Imperfecti, *Cenococcum graniforme*, is often present. The anatomical features of ectotrophic mycorrhizas have been described systematically (Dominik, 1959; Chilvers and Prior, 1965) and arranged in a series of subtypes. Such work will assist in sorting out mycorrhizas before identifying the fungi concerned, but there is no guarantee that different anatomical types represent different species of fungi, or that one species of fungus may not take various anatomical forms. The problem can best be solved by inducing the fungi to form fruitbodies in culture, though this is no easy task.

In contrast to the ectotrophic mycorrhizas there are other types termed **endotrophic**, whose fungi develop from a loose weft of hyphae exterior to the root, without ensheathing it, and whose hyphal branches enter the root cortex and penetrate cortical cells. Endotrophic mycorrhizas are to be found in a very wide variety of plants, but have been investigated especially in the Ericaceae, Orchidaceae, Bryophyta and various grasses and cereals. It has been found that some have a regularly septate mycelium characteristic of higher fungi, while in others the mycelium is described as 'non-septate' and assigned to members of the lower fungi. Among the higher fungi that have been identified are species of the genera *Marasmius* and *Phoma*, *Armillariella mellea*, and several species of the form-genus *Rhizoctonia*. The perfect states of these rhizoctonias have so far been found to be species of the genera *Thanatephorus*, *Ceratobasidium*, *Tulasnella* and *Sebacina*. It is not uncommon for the intracellular hyphae to become coiled as **pelotons**, and for them to be digested by the root cells.

The type of endotrophic mycorrhiza formed by lower fungi is usually known as a **vesicular-arbuscular mycorrhiza**. The main hyphae, which pass between cortical cells, are relatively wide and mostly without septa, but are able to form septa in some circumstances. At intervals they bear intercalary or apical swellings (**vesicles**) which are large and thick-walled. Smaller branches penetrate the cells and there become more or less dichotomously branched into bushy **arbuscules**, a type of haustorium, whose branchlets may be swollen at the ends. Fossilized vesicular-arbuscular mycorrhizas have been found in the Carboniferous and Devonian strata. Their modern counterparts have been investigated, especially in the Gramineae. Two types have been identified, although both were formerly included in the genus *Rhizophagus*: they are a species of *Pythium* (Mastigomycotina) and certain other fungi which are almost certainly imperfect Endogonaceae (Zygomycotina) and possibly small forms of *Endogone*.

MYCANGIA

In some groups of insects, especially wood-boring (e.g. *Xyleborus*) and bark-feeding (e.g. *Ips*) beetles, and woodwasps (e.g. *Sirex*), the adult females have special pouches called mycangia (Batra, 1963, 1966, 1967; Francke-Grosmann, 1967) containing a fungal inoculum which the insect deposits in wood during oviposition. Within a mycangium, the fungus usually assumes a short-celled arthrosporic, blastosporic or yeast-like form, but it is mycelial in form when it grows in wood or in culture. The fungus grows in the insect tunnels and in nearby wood. In some instances the fungus itself is a major source of food for the developing larvae; in others it may simply soften the wood and assist the larvae to tunnel more easily. Most of the fungi thus carried by insects are saprobes which do little damage apart from causing blue-staining of the wood; but a few are parasitic and have pathogenic effects on the trees. These fungi, often known as 'ambrosia fungi' are mostly Ascomycotina (particularly Endomycetales and Microascales; genera *Leptographium, Ceratocystis*) or Fungi Imperfecti, but a few are Basidiomycotina (particularly species of *Amylostereum* associated with Siricid woodwasps).

TAXONOMIC IMPLICATIONS OF SOMATIC STRUCTURES

The thallus

If fungi are divided into those types whose thalli are motile amoeboid plasmodia lacking a true cell wall, and those whose thalli are bounded by true walls, two groups are obtained corresponding roughly with the Divisions Myxomycota and Eumycota, respectively. It is true that there are some intermediate types whose taxonomic position is doubtful: at least in early development the holocarpic, parasitic thalli of members of the Plasmodiophoromycetes and some of the Chytridiomycetes (Olpidiaceae, Synchytriaceae) are amoeboid, and do not form mycelia. In classifying these, however, we rely not merely on the single character of the thallus, but also on many other associated characters, since reliance on a single character is always suspect in taxonomy. The general features associated with the plasmodial thallus of Myxomycota (Chapter 8) combine to set this Division apart from other fungi as a well defined group. But the gulf between Myxomycota and Eumycota is not as wide as it might appear at first sight. First, intermediate types exist among some of the Chytridiomycetes, as indicated above. Second, we can conceive of a unicellular or mycelial thallus of the Eumycota as having evolved from an amoeboid type by thickening and change in the hyaloplasm membrane to become a true wall, without necessarily implying that Eumycota are descended directly from Myxomycota.

With the exception that absence of chlorophyll carries wide physiological implications, the usual definition of a fungus (p. 15) succeeds in giving only a morphological picture; it carries no reference to the extraordinary and meaningful behaviour of fungal protoplasm moving within a system of tubes. This is admirably conveyed by Langeron's (1945) definition of a fungus, here translated: 'A fungus is a nucleated cytoplasmic mass which displaces itself centrifugally, sometimes occurring free, sometimes in the interior of tubes which it constructs about itself as it progresses peripherally. As the tubes elongate under cytoplasmic pressure the fungus quits the central part of the thallus, little by little, as these elements become progressively vacuolated, empty themselves and die. This sort of fungal thallus, whatever its form, always comprises a living peripheral zone with continuous growth and a central skeletal zone formed of empty and dead tubes. Growth of the thallus is limited only by exhaustion of the cytoplasmic mass, which ends by passing entirely into propagative or reproductive spores.'

Gregory (1966) questions the hypothesis that vacuolar pressure causes cytoplasmic flow, but emphasizes that the 'essential point is that the fungus moves in the tubes until the substrate has been exploited enough, then it quits the tubes where it vegetated, and launches out in the form of resting spores or dispersal spores (or it may form sclerotia . . .). The spores now become the essential fungus, as distinct from the dead and more or less empty tubes, which may be all that remain for examination in the herbarium.'[1] Gregory's view of a fungus as a 'plasmodium moving about in a system of tubes' does not, as he points out, 'necessarily imply descent from the Myxomycetes; nor does it exclude it.'

Two further facts are interesting to note in this regard. First, it has been shown in work reviewed by Robertson (1965) that if partly plasmolysed hyphae are treated with snail enzyme the hyphal walls are destroyed and rounded protoplasts are released; the latter are bounded by a pellicular membrane and are able to regenerate wall material and revert to a hyphal construction. Second, Martin (1932) has noted that some of the principal 'veins' in a Myxomycete plasmodium are enclosed by hyaloplasm membranes and can be lifted out intact from the matrix in the form of tubules, which he compares with the empty tubes left behind by certain rapidly growing hyphae of mycelial fungi. He states that the membrane present 'appears to differ in degree, rather than in kind, from that surrounding the protoplasm of a typical Phycomycete'. While these facts do not permit us to state with any certainty that hyphal walls have evolved from hyaloplasm coverings of plasmodia, they are nevertheless suggestive.

Among the Eumycota, or non-plasmodial groups of fungi, unicellular, pseudomycelial and mycelial thalli are encountered, but the distribution of these types is too widespread and irregular for use in delimiting major taxa; and in species with dimorphic thalli they may occur in the same species. Unicellular thalli are, however, notable among the Chytridiomycetes and, together with pseudomycelial thalli, are especially characteristic of the yeasts

[1] In conformity with this view of a fungus, Gregory describes a fungal spore as follows: 'The fungus spore, in contrast to the vegetative mycelium, is a nucleate portion delimited from the thallus, characterised by cessation of cytoplasmic movement, small water content and slow metabolism, lack of vacuoles, and specialised for dispersal, reproduction or survival.' The various types of spore are discussed later in this book.

(Ascomycotina, Endomycetales) and the 'mirror yeasts' or Sporobolomycetaceae (Basidiomycotina).

Mycelia and colonies

Although individual hyphae can be described in terms of colour, width, wall-thickness, septation and average cell-length, type and frequency of branching, it is an extremely difficult task to describe a mycelium. Its gross appearance may be given various general terms (such as woolly, cottony, silky, fibrous, felted, floccose, fluffy, mealy etc.) but these are rather vague and each might apply to many different types of fungi. It is difficult, too, to relate gross appearance to the microstructure responsible for producing a particular type of texture. We can describe some of the more unusual types of microstructure but are flummoxed by the commonplace. This is an area which requires more investigation and thought. The work of Nobles (1948, 1958) with polypore mycelia in culture and that of Pantidou (1961) with boleti, are notable examples of the successful use of mycelial characters in taxonomy. Some sterile mycelia are reasonably distinctive, but others are not. Even when two cultures are directly compared it requires great experience of culture work to be able to say with some assurance that they are likely (or unlikely) to represent the same species, unless the mycelium can be induced to form some type of fructification or spores.

Colonies, too, are difficult to describe. The yeast colony is usually distinctive; it spreads slowly and forms a raised, moist, glutinous mass which often becomes wrinkled; but some fungi other than yeasts can assume a yeast-like form. The colonies of moulds (Deuteromycotina and Mucorales) can usually be recognized by their abundant spores produced mostly on simple sporophores. In general, however, mycelial and colonial characters are not useful for characterizing major taxa of fungi; their greatest application is at the species level, and even there are often not completely reliable.

Hyphae

In practice it may be difficult to determine the type of septum and septal pore, or to determine nuclear numbers and the relationship of nuclear division to septum-formation, without the use of special techniques and laborious observation. However, this does not invalidate the use of hyphal characters for taxonomic purposes in those fungi which have a mycelial thallus.

The Eumycota are traditionally divided into two large groups commonly called the 'lower' and 'higher' fungi. Because these groups are heterogeneous they have no official status in classification, but they are nevertheless a most useful concept in discussions of the Eumycota.

It is conventional in mycology to insist that lower fungi ('Phycomycetes', i.e. Mastigomycotina and Zygomycotina) 'have coenocytic hyphae' and this leads to the quite untrue statement that their hyphae are characterized by 'having no septa' (sometimes adding, 'at least when young'), whereas the hyphae of higher fungi (Ascomycotina, Deuteromycotina, Basidiomycotina) are 'non-coenocytic and septate'. Alternatively, to insist that hyphae of lower fungi can be recognized by their multinucleate cells is not always true (e.g. uninucleate in *Basidiobolus ranarum*) and ignores the fact that cells of higher fungi may also be multinucleate in some species and in some circumstances. In most lower fungi, however, the young and actively assimilating thallus is a coenocyte, since it is a single multinucleate cell (even when branched) in which nuclei divide without this process being followed directly by cytoplasmic cleavage due to septation. But this state does not persist. In older thalli there are always some adventitious septa which not only cut off exhausted parts, but also are found with some frequency in developing parts, especially the sporophores. The whole thallus is no longer a single cell; and its individual cells, though coenocytic, are effectively isolated from one another by plugged septa. In contrast, the hyphae of a higher fungus are usually divided by incomplete primary septa formed in association with nuclear division. The cells so formed may be uninucleate, binucleate or multinucleate according to species and to various features of nuclear behaviour, but they communicate directly through open septal pores. While it is true that such cells are usually reduced in nuclear number and appear able to function independently, it could be held that the septa do not divide the cytoplasm completely and that the whole hypha or thallus is a single multinucleate cell. Are such hyphae, then, coenocytic or non-coenocytic? A useful primary grouping of Eumycota into 'lower' and 'higher' fungi can be made on hyphal characters, but only if stress is laid on the origin, type and relative frequency of septation rather than on whether the hyphae are 'septate or non-septate', 'coenocytic or non-coenocytic', or 'coenocytic or cellular'.

The mycelium of most lower fungi can be recognized in practice by its often, but not always, wide hyphae with irregular, comparatively sparse septation associated with cytoplasmic accumulation or withdrawal (which is usually readily detected with ordinary mycological stains), and by reproductive structures. Absence of anastomosed hyphae also points to the lower fungi. Mycelia of higher fungi are recognized by their usually, but not always, narrow hyphae with regular primary septation in acropetal succession from hyphal apices, and by reproductive structures. Anastomosis is a common feature of hyphae of higher fungi. The nature of the septation in each group is best determined in young marginal parts of the thallus, by working back from the apices of the hyphae. Some types of reproductive structures, such as the various types of thallospores, may be common to lower and higher fungi, but there are others that are quite distinctive.

In some instances it is possible to distinguish sterile mycelia of Ascomycotina from those of Basidiomycotina using the light microscope and suitable stains.[1] The simple septal pore of Ascomycotina, and the dolipore of Basidiomycotina can often be distinguished in this way provided that they are big enough in the species under examination. Clamp connexions (p. 195) are easily seen and, when present, are indicative of Basidiomycotina.

Primary septa, as far as is known, predominate in the higher fungi and it is interesting to note that the presence of regular septation in these essentially non-aquatic fungi must undoubtedly strengthen their hyphae and contribute to their ability to form fruiting structures which are raised above the substratum, thus increasing the efficiency of spore-dispersal by agents other than water. The common occurrence of anastomosis in hyphae of higher fungi must also increase their mechanical strength as well as consolidate their tissues.

Adventitious septa predominate in the lower fungi. Their strengthening function in the mycelium of lower fungi would appear to have little importance since they are relatively sparse and since many of these fungi inhabit water, soil moisture films, or are parasitic in the cells of plants or animals. In all these situations the sporophores are comparatively small, are apparently adequately supported, and are associated with spore types and spore-dispersal mechanisms which do not require massive raised fruitbodies for efficient spore dispersal. However, such septa usually occur in the erect simple sporophores of these fungi and, together with turgor, may assist in keeping them erect.

Specialized somatic structures

All the specialized somatic structures described above may have great taxonomic value in the differentiation of species, but are of variable significance at higher levels of classification. The parasitic habit is widespread among fungi and unsuitable as a character in classification except in certain groups of fungi which show other close relationships. Any structures associated with parasitism, such as haustoria, appressoria or hyphopodia, have limited taxonomic value, their use being confined mainly to specific delimitation. Rhizomorphs and mycelial strands are of some assistance in the taxonomy of sterile mycelia since they appear to be confined to the higher fungi and are more common in Basidiomycotina than in Ascomycotina. Rhizoids, on the other hand, are confined to the lower fungi, and rhizomycelia to the Chytridiomycetes. Complex stromata are found almost exclusively in the higher fungi, where compact fruitbodies are common. Stromata are widespread in Ascomycotina and exist in several distinctive types, thus providing excellent characters for some of the major groupings of Ascomycotina.

[1] In my opinion the most useful stain for temporary mounts of fresh or dried fungi is ammoniacal congo red (Boidin, 1958), which stains fungal walls delicately but precisely. It also stains cytoplasm, usually sufficiently, but if this needs to be intensified the congo red can be mixed with a small trace of 2% aqueous phloxine on the slide at the time of mounting.

Stromatic fruitbodies are extremely rare in lower fungi, a notable example being in the genus *Endogone*, some of whose species have underground fruitbodies which are compact and stromatic. Sclerotia are widely distributed among higher fungi, but have generally been considered too difficult to characterize for taxonomic use. However, in the genus *Typhula*, Remsberg (1940) made much use of sclerotial anatomy for distinguishing species in the absence of their fructifications, while in the Sclerotiniaceae Whetzel (1945) took their gross anatomy into account in generic diagnoses. Townsend and Willetts (1954) observed that some species assigned to *Sclerotinia* show different types of sclerotium-development and suggested that taxonomic use could be made of this fact.

Summing up some of the taxonomic implications of characters of the thallus and hyphae, we can speculate that in the fungi there has been a broad phylogenetic trend from a truly coenocytic condition towards one where the thallus is divided into virtually self-contained cells. The plasmodium of a slime-mould (Myxomycota) is a cell bounded only by a hyaloplasm membrane and may be a true coenocyte (as defined above) or a multinuclear mass of protoplasm formed by fusion of several protoplasts without fusion of their individual nuclei. The hyphae of lower fungi (Mastigomycotina and Zygomycotina) are true coenocytes when young, but later tend to develop a limited organization of separate coenocytic cells. The higher fungi (Ascomycotina, Basidiomycotina and Deuteromycotina) have hyphae divided into cells which probably act for the most part as separate functional units, but still retain communication through their septal pores. These types of structures seem to show a sequence of evolutionary trends, although the fungi bearing them are not necessarily descended directly one from the other.

5
REPRODUCTION IN FUNGI: GENERAL

Most fungi are able to reproduce both asexually and sexually, and both methods usually imply the formation of spores from various types of sporogenous cells borne on either simple or compound sporophores. However, in asexual reproduction, hyphal cells are sometimes converted directly into spores; or stromatic groups of hyphae may have a propagative function, as with sclerotia. A third type of reproduction is that known as parasexual reproduction.

ASEXUAL, SEXUAL AND PARASEXUAL REPRODUCTION CONTRASTED

The hereditary properties which determine the expression of particular characters in individuals are carried by **genes** at loci on **chromosomes** within the cell nuclei. The type of reproduction depends primarily upon the type of nuclear behaviour.

Asexual

The nuclei of somatic cells in some lower fungi are able to divide by constriction independently of cell division; little is known about the genetic effects of this process except that the chromosomes are rearranged in an orderly manner in the daughter nuclei. In general, however, it may be said that when a somatic cell divides, its nucleus divides by true **mitosis** into two daughter nuclei each of which has identical chromosomes formed by longitudinal halving of the chromosomes of the original nucleus. The daughter cells thus have the same genetic constitution as the original cell and, if one or both are liberated — perhaps as spores — and give rise to new individuals, asexual reproduction has occurred

Sexual

With sexual reproduction the essential feature is a union of a pair of unlike haploid nuclei derived from different parent cells, and reassortment of the genes during subsequent **meiosis** of the diploid nucleus so that the two resulting haploid nuclei differ in genetic constitution from the parent nuclei and some variation from the parent types occurs in the offspring.

The haploid nuclei of all individuals of a species carry similar sets of chromosomes characteristic for the species, and each chromosome carries similar sets of genes. Any one type of gene, however, can exist in a number of different forms called **alleles**, each concerned with the expression of a particular character determined by the gene type, but differing in the extent to which they bring about its determination. Thus any pair of mating individuals of the same species have the same genes, but in different allelic form, at the same chromosomal loci. Mating occurs by some form of cell anastomosis (**plasmogamy**; see p. 91) which brings haploid nuclei of opposite mating type into a single cell, where they fuse (**karyogamy**) and form a diploid nucleus which contains two complete sets of chromosomes characteristic of the species, one set being derived from the nucleus of each parent. On meiosis of the diploid nucleus, the chromosomes do not split but instead are segregated whole into two complete sets, each of which forms the chromosome complement of a haploid daughter nucleus. However, some of the chromosomes in each daughter nucleus have now been derived from the one parent, and others from the other parent, and consequently the daughter nuclei now carry a reassortment of alleles (**recombination**). Any progeny of the daughter cells, which may be spores, will therefore have the general characteristics of the species, but will not be genetically identical with either of the parent types.

Parasexual

Recombination of the hereditary properties in fungi may be either sexual or parasexual. In a sexual cycle the processes of plasmogamy, karyogamy and meiosis take place in a regular sequence and at specific stages in the cycle. In the parasexual cycle, a phenomenon discovered by Pontecorvo and Roper (1952), recombination of hereditary properties is based on crossing-over during the mitotic instead of the meiotic (sexual) cycle.

Parasexual recombination takes place as follows. The mycelium becomes heterocaryotic by anastomosis or mutation. A proportion of haploid nuclei in the heterocaryon fuse to form diploid nuclei, which divide mitotically and increase in number together with the remaining haploid nuclei. At mitosis of some of the diploid nuclei, recombination by crossing-over occurs. Some of the diploid nuclei may now enter spores which germinate and produce homocaryotic diploid mycelia. The diploid nuclei in these mycelia become haploidized, but the resulting haploid nuclei differ in genetic constitution from those of the original heterocaryotic mycelium because of mitotic recombination having occurred. The various processes may all be occurring simultaneously in the same thallus, not in a regular sequence or at specific stages. The parasexual cycle may or may not be accompanied by a sexual cycle. Parasexuality may clearly be an important mechanism of genetic variation in those fungi which lack the capacity for sexual reproduction, or in which it is rare.

SPOROPHORES

A sporophore is a specialized branch or tissue arising from the thallus and bearing sporogenous cells and spores. More than one type of sporophore may be associated with a single thallus, or with different mycelia of a single species. Thus in a single thallus, or species, various types of spores may characterize different phases of development. The phase associated with asexual spores or sterile mycelia is known as the **imperfect state** of the fungus, while that associated with production of zygotes or of spores resulting from any type of sexual process is the **perfect state.**

Growth of sporophores is always more limited than that of somatic hyphae and is either terminated or slowed down on production of spores. Limitation of growth is usually accompanied by the formation of characteristic structures which are sharply distinct from somatic hyphae and thus often diagnostic of various taxa. The sporophores are often the most conspicuous part of the fungus and may even be mistaken for the whole fungus.

Simple or filamentous sporophores

These are branches of the somatic hyphae which give rise to sporogenous cells and spores on reaching a length which is roughly determinate for a particular species. The sporophores are usually erect and sometimes distinctively branched. Good examples of simple sporophores are to be found in the lower fungi (as sporangiophores) and in the Deuteromycotina (as conidiophores). Growth of the sporophore may cease with the formation of sporogenous cells or spores, but in some types may be resumed again by proliferation to form further cells. In general it may be said that although the sporogenous cells and proliferative methods may appear to be highly diversified they are, in fact, variations of a few basic types which are determined by the properties of hyphae, from which they arise. (Compare Chapter 13.)

Compound sporophores

These are stromatic or semistromatic structures formed by concerted growth of closely associated hyphae, and they contain or bear layers of aggregated sporogenous cells and spores; they comprise the fructifications or fruitbodies of the larger types of fungi and assume a limited number of characteristic forms, e.g. flattened layers, clubs, cushions, cups, spheres, brackets and stalked pilei. The tissues of the fruitbody give protection and support to the sporogenous cells and generally raise them above the substratum in a manner promoting efficient discharge and dispersal of spores. Indeed, the type of fructification is always closely linked with its function as a bearer and discharger of spores and with the general biology of the species. Cupulate fruitbodies, for example, are able to hold water if they face upwards; in the cupulate Ascomycotina (Discomycetes) this is associated with spore-discharge by bursting of turgid cells or by swelling of a mucilaginous matrix. On the other hand, in the Basidiomycotina the spore-discharge mechanism will not work if the sporogenous cells (basidia) are wet, and the fruitbody usually faces downwards so that the basidia are protected from free water. The cyphelloid Basidiomycotina, with cupulate fruitbodies, may look very much like Discomycetes, but observation of the direction in which the fruitbody faces, whether upward or downward, should enable one to distinguish them in most cases without resort to microscopy.

Sporogenous cells may be aggregated in a layer lining the inside of various types of cavity in the fruitbody (e.g. pycnidia, perithecia) or in a layer beneath the epidermis of a host plant (e.g. acervuli, sori), or in a layer which is external to the bulk of the compound sporophore. In Ascomycotina and Basidiomycotina the sporogenous layers comprising asci or basidia, respectively, and any associated sterile structures, are known as the **hymenium** (p. **hymenia**).

SPORES

In mycology the term spore is applied to any small propagative, reproductive or survival unit which separates from a hypha or sporogenous cell and can grow independently into a new individual; such a spore is composed of either a single cell or a relatively small number of cells joined together in a unit which is abstricted and dispersed as a whole. Another, complementary, view of a spore is given earlier on p. 70.

Spores contain one or more nuclei derived either from nuclear fusion or from nuclear division; thus spore-formation is closely connected with nuclear behaviour. In the higher cryptogams (Bryophyta and Pteridophyta) there is a distinct alternation of the haploid, gametophytic generation with a diploid, sporophytic generation, and the term spore is applied to units formed after meiosis in the sporophyte; such spores are therefore **meiospores**. In fungi, however, there is no differentiation into sporothalli and gametothalli (except uniquely in the genus *Allomyces*), and meiospores of various types (e.g. ascospores, basidiospores, sporangiospores of Myxomycetes) may be produced from the same thalli as **mitospores**; the latter do not result from meiosis but instead are formed after mitotic division of the nuclei (e.g. zoospores, conidia, uredospores).

In most of the lower fungi, the **zygote** resulting from sexual fusion has the form and function of a spore and is generally known as a sexual spore (e.g. oospore, zygospore). In higher fungi, however, the zygote is a diploid nucleus produced in a cell which then gives rise to a number of spores (ascospores or basidiospores). These are commonly known as 'sexual spores' because of their association with karyogamy and meiosis, but they are not

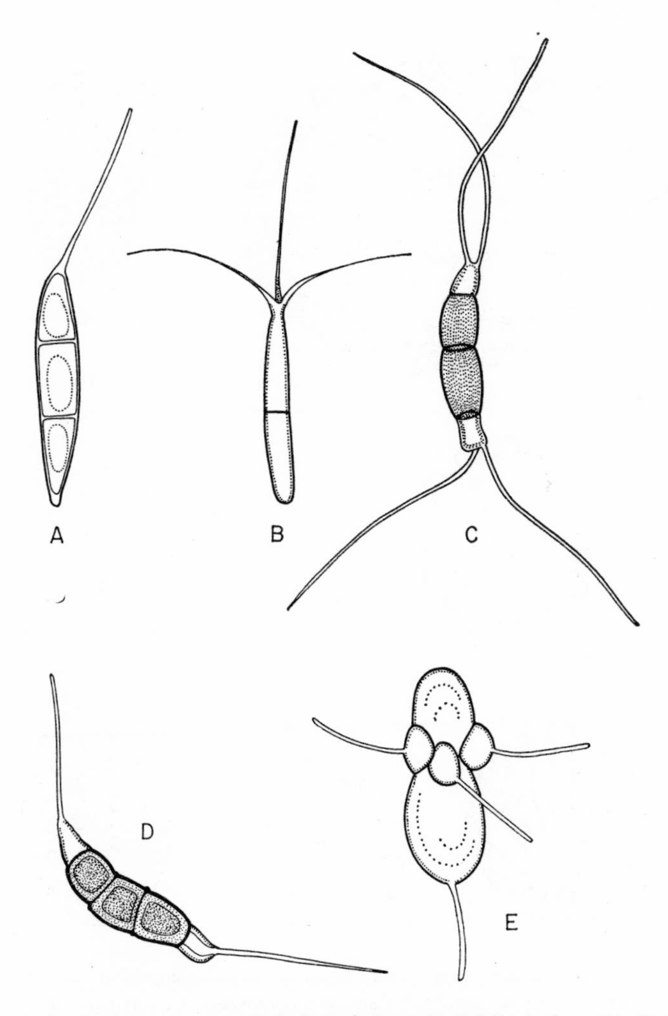

Fig. 15. Appendages on spores of: A, *Pyricularia* sp.; B, *Robillarda phragmitis*; C, *Diploceras* sp.; D, *Cryptostictis* sp.; E, *Entomosporium maculatum*.

homologous with the sexual spores of lower fungi. Because of this they are sometimes distinguished as 'perfect' spores, being produced by the 'perfect state' of the fungus.

That the term spore is capable of wide interpretation must be obvious. It is a general term which, in mycology, is qualified by particular prefixes to indicate numerous special types. For example, an ascospore is a spore produced in an ascus (typification by origin); the same ascospore may also be called a didymospore or described as didymosporous if it is two-celled (typification by morphology); and further it may be classed as a xenospore

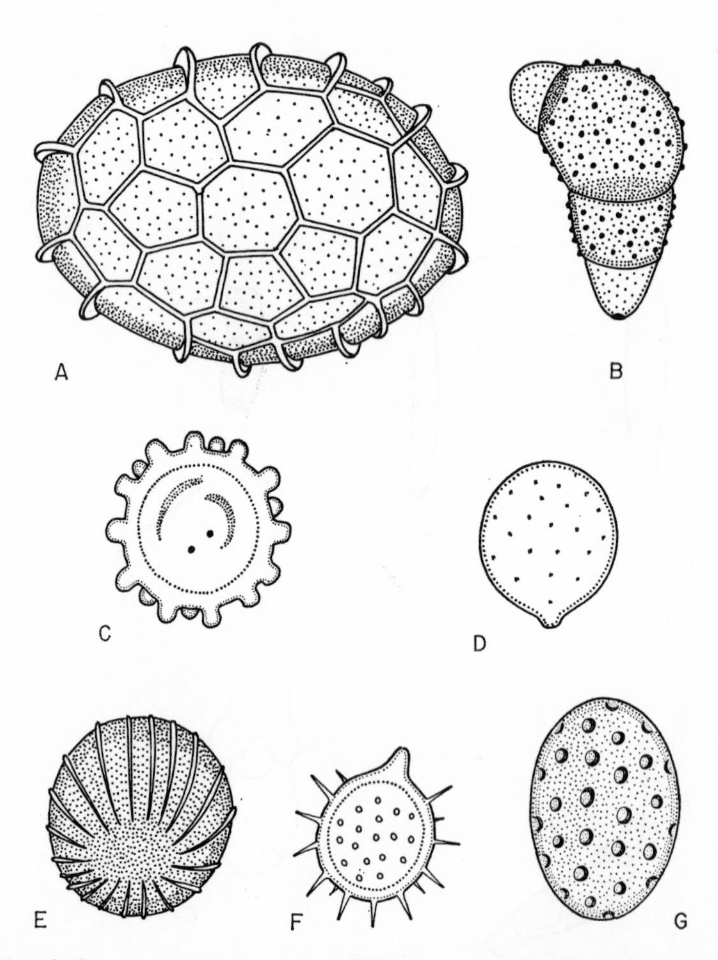

Fig. 16. Spore ornamentation (*camera lucida* at various magnifications). A, reticulate (*Tuber clarei*); B, verrucose (*Curvularia verruculosa*); C, verrucose (*Entorrhiza calospora*); D, punctate (*Peniophora sphaerospora*); E, striate (*Uromycladium tepperianum*); F, echinate (*Botryohypochnus isabellinus*); G, alveolate (*Gelasinospora* sp.).

if it is a dispersal spore or as a memnospore (Gregory, 1966) if it is a sedentary spore (typing by biological function). At first sight this may appear confusing, but in practice it is useful and easy to comprehend, and to apply according to circumstances.

Spores are commonly unicellular but in many species they may be divided by septa into a number of cells; the septa are commonly transverse, but again in many fungi they may be both transverse and longitudinal together. Each cell of a multicellular spore may be uni-, bi- or multinucleate according to the species of fungus. Various spores may be motile or non-motile, thick-walled or thin-walled, hyaline or coloured, smooth or with ornamented walls, and with or without simple or branched filiform appendages (Fig. 15). The chief types of wall ornamentation (Fig. 16) recognized for taxonomic purposes are the following: **echinate**, with sharp spines; **verrucose**, with small rounded or truncate warts; **reticulate**, with anastomosed ridges or plates forming a surface network; **striate**, with lines, grooves or ridges running roughly parallel to one another; **alveolate**, with small superficial pits giving a poroid appearance; and **punctate**, with minute markings whose exact nature cannot be resolved with the light microscope. The terminology used to describe some of the commoner shapes of spores is illustrated in Fig. 17.

The great variety in form of spores lends itself to taxonomic use; consequently, it is convenient to have a terminology which describes any kind of spore according to its colour and septation. Such a terminology, set out below, was devised by Saccardo (1886) and extensively used in his classification of Deuteromycotina. But it should be stressed that the terms are primarily

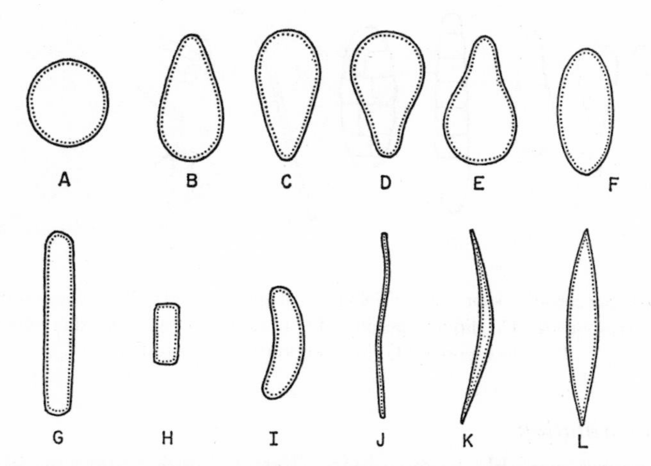

Fig. 17. Spore shapes (diagrammatic). A, spherical; B, ovate; C, obovate; D, pyriform; E, obpyriform; F, ellipsoid; G, cylindrical; H, oblong; I, allantoid; J, filiform or scolecoid; K, falcate; L, fusoid.

descriptive, can be applied to spores of any other taxonomic group, and do not indicate types of spores differentiated by their origin or by the class of fungus to which they belong. Indeed, it is found in many species that the colour and septation of the spores may be quite variable, and change with maturity. Hawker (1957) has pointed out that all spores are originally single-celled, colourless and globose to ovoid; subsequent changes in shape and ornamentation are brought about by the formative effects of equal or unequal pressures in the immediate environment of the developing spore, by the shape of spaces which it can fill, and by differences in the rates of expansion or hardening of walls. Development of septa is also a gradual process. In all these respects the spore behaves like the apex of a hypha.

Saccardo's spore terminology (Fig. 18)

Allantospore — unicellular, with rounded ends and a curved outline (sausage-shaped).
Amerospore — unicellular and short in proportion to breadth.
Didymospore — two-celled.
Phragmospore — with two or more transverse septa.
Dictyospore — muriform, i.e. with both transverse and longitudinal septa.
Scolecospore — long and filiform, many times as long as they are broad, with or without transverse septa.
Staurospore — stellate, having three or more arms.
Helicospore — coiled in two or three dimensions, forming a flat or spiral coil.

To each of these terms a prefix can be added to indicate the colour of the spore: **hyalo-**, for hyaline spores; **phaeo-**, for spores with yellow, brown or darker walls.

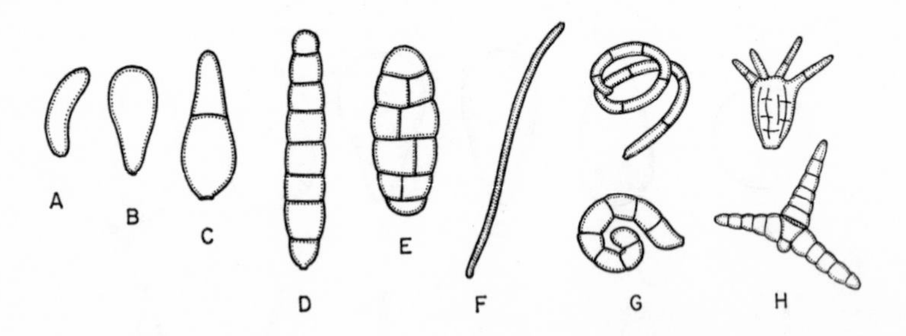

Fig. 18. Saccardo's spore terminology (diagrammatic). A, allantosporous; B, amerosporous; C, didymosporous; D, phragmosporous; E, dictyosporous; F, scolecosporous; G, helicosporous; H, staurosporous.

Spore germination

Some spores are able to germinate (Fig. 1) immediately on liberation; others undergo a period of dormancy which is broken only under special conditions. For example, spores of *Anthracobia melaloma*, a Discomycete, have their dormancy broken by heat and germinate in burnt soil after a fire which would destroy most other organisms. The advantages of dormancy are obvious: the spores germinate only when conditions are favourable for them and possibly unfavourable to other, competing organisms.

The majority of spores germinate by means of a **germtube**. In spores investigated by electron microscopy the wall is usually two-layered at least, and during germination the inner wall protrudes through a weakened area of the outer wall as a germtube initial. The germtube then elongates into a typical hypha. Each spore may put out one or more germtubes. Each cell of a multicellular spore may be capable of germinating, and in effect such a spore is a group of unicellular propagules, but is usually regarded as a unit because it is dispersed as a whole.

Resting zygotes (oospores, zygospores and resting sporangia) tend to vary in their behaviour during germination. Some differentiate sporangiospores (zoospores or aplanospores) internally and these germinate by a germtube on liberation; others put out a short tube with an apical sporangium containing sporangiospores; still others may germinate directly by means of a germtube.

Zoospores in some members of the Saprolegniales germinate by a germtube only after alternating periods of motility and encystment. The primary zoospore differs in morphology from the secondary and subsequent orders of zoospores liberated from the cyst-cells.

Some types of spores which are forcibly discharged (e.g. in the Entomophthorales and some Basidiomycotina), germinate by a process known as **repetition** (Fig. 19). Here the spore puts out a short spicule, or secondary sterigma, which bears a secondary spore morphologically similar to, but usually smaller than, the primary spore. The secondary spore is then liberated and germinates normally by means of a germtube, though sometimes successive repetition may occur before a spore is formed which finally germinates by a germtube. Taken in conjunction with other taxonomic

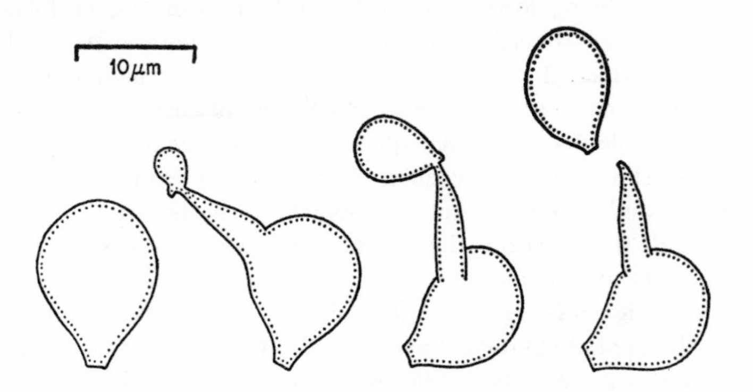

Fig. 19. Germination by repetition (basidiospores of *Thanatephorus orchidicola*).

features, the character of spore repetition has proved useful in the generic delimitation of some Basidiomycotina, even though repetition may be inconstant and depends in one investigated species (*Thanatephorus cucumeris*; Whitney, 1964) upon the presence of certain ether-soluble substances which are erratically produced. The biological advantage of spore repetition must surely be that it gives a second chance to spores that have fallen back on the hymenium or substratum, allowing them to produce further spores which can be discharged into the airstream above the substratum and so be efficiently dispersed. Repetition should not be confused with the production of numerous small conidia (blastospores) from basidiospores (e.g. in members of the Dacrymycetaceae) or from ascospores (e.g. in *Taphrina* and *Coryne*).

PLEOMORPHISM

The existence of fungal species in more than one independent morphological form, with different types of spores typifying the different forms, is known as **pleomorphism**, and a species is said to exist in a number of **pleomorphic states.**

In the lower fungi, asexual and sexual states are mostly formed in sequence on the same thallus and it is usually possible to observe direct continuity between particular states of a species; in any case, the asexual states alone are sufficiently distinctive to form a large part of the classification of these fungi. In the higher fungi, however, the asexual (mycelial or conidial) states may exist quite independently of the sexual states and be completely different in morphology. Although the mycelia giving rise to sexual fructifications and asexual sporophores, respectively, may be similar in a single species, they are not necessarily so and mycelial characters alone are rarely distinctive enough to be diagnostic and to permit the various states of the same species to be connected.

The real or apparent independence of the sexual and asexual states may be due to their being formed on different host substrata, or formed at different seasons so that they are unlikely to be noticed and collected together. Sometimes the sexual state is formed only under specialized conditions, genetic or environmental, which are unknown to us; but sometimes it is probable that particular species are incapable of forming a sexual state at all. It should also be recalled that several different types of asexual fructifications and asexual spores may sometimes be produced by a single species. For all these reasons the genetic continuity of sexual and asexual states may not be at all obvious.

Among the higher fungi encountered in an 'imperfect state' it is clear, both from the hyphal characters and from the types of asexual spores, that one is dealing with a member of the Ascomycotina or Basidiomycotina and not with a member of the lower fungi. However, apart from the characters of septal pores which may be difficult to observe, there is usually insufficient information to enable one definitely to assign specimens to one or other of these Subdivisions. Such assignation would depend upon seeing asci, which characterize the Ascomycotina, or seeing basidia or clamp connexions of the Basidiomycotina; that is, these Subdivisions have to be recognized and classified primarily according to their sexual states. It is not possible to delimit the Ascomycotina and Basidiomycotina on the basis of their asexual spores although, with hindsight, it is known that the majority of conidial states belong to the Ascomycotina. This residue of sterile or conidial fungi with the mycelial features of higher fungi includes many of the most important species as regards pathogenicity and ability to spoil or destroy materials of use to man, and also many species which produce valuable metabolic products. Thus it is a matter of practical necessity and convenience to be able to name and classify them in the state in which they happen to be found. The only reasonable solution is to create a taxon to accommodate such imperfect states. This taxon is the artificial Subdivision known as the Deuteromycotina, or Fungi Imperfecti.

Effect of pleomorphism on nomenclature

In Chapter 2 it was pointed out that it is permissible to use different names for the different states of a species with a pleomorphic life-cycle. While one may possibly wish to refer to the mycelial *Rhizoctonia rubiginosa* and conidial *Oedocephalum glomerulosum* states of *Ascophanus carneus* (Warcup and Talbot, 1966), this multiplicity of names for the same species can be carried to extremes. Mason (1937) has quoted a species described by Briosi and Farneti under seven different binomials simultaneously, each referring to a different morphological state of the fungus and reflecting its developmental variability. In general, the fact is that one tends to study the anatomical parts of a fungus and not the whole organism in all its phases, but, as Mason (1940) has observed, it is difficult to do otherwise because an individual fungus is elusive; one does not always know where one individual ends and another begins.

6
ASEXUAL REPRODUCTION

Asexual reproduction does not involve the union of sex organs (**gametangia**), sex cells (**gametes**) or nuclei. Any asexual spores formed are mitospores, not zygotes or meiospores. Any of the following processes may be concerned in asexual reproduction.

TYPES OF ASEXUAL REPRODUCTION

Fission of Unicellular Thalli

In this process, the nucleus of a somatic cell divides mitotically and the cell contents become divided transversely by a septum into approximately equal parts. Separation of the two daughter cells occurs at the septum and they round off into cells of similar size, shape and contents, becoming two separate thalli. Fission is typical of some types of yeasts.

Fragmentation of the Mycelium

Small fragments of mycelium taken from most parts of a living thallus and including undamaged cells, are usually able to grow into a new thallus if the right conditions are provided. This property of fungi is used extensively by mycologists in making laboratory cultures; but it should also be remembered that breaking of a mycelium can take place under natural stresses, for example in turbulent water. A mycelium in soil may be subjected to conditions resulting in the death of most of the hyphae, but small pieces may survive, possible within fragments of organic material, and are important in recolonizing the soil saprobically and perhaps spreading to penetrate and parasitize living parts of plants.

Production of Asexual Spores

Virtually any short cell of a thallus has the potential to become an asexual spore if it is capable of being set free and existing independently. Apart from this, however, sporophores of various types may be formed from the thallus and give rise to new cells which act as spores. According to their methods of formation the types of asexual spores can be divided into two main categories: sporangiospores and conidia.

Sporangiospores

In the formation of sporangiospores (Fig. 39), a sporophore (in this case a **sporangiophore**) becomes swollen, commonly apically but sometimes in an intercalary position, to form a sac-like **sporangium.** This structure is bounded by a membranous wall (**peridium**) derived from the sporangiophore, and contains protoplasm with a large number of haploid nuclei. The protoplasm cleaves into uninucleate segments each of which rounds off and becomes transformed into a large number of spores. Thus the sporangiospores are asexual spores formed within a sporangium.

Motile sporangiospores are known as **zoospores**; they are 'naked', i.e. they have no cell wall. Non-motile ones are called **aplanospores** and have walls secreted about them during their formation. In some of the lower fungi, the sporangia and zoospores are indistinguishable except in function from the gametangia and gametes, respectively.

Conidia

The term conidium (s; Figs. 43–50) is generally used for any asexual spore except a sporangiospore or (in some definitions) an intercalary chlamydospore. It is desirable sometimes to distinguish between two main types of conidium, namely thallospores and 'true conidia'. When this distinction is made the latter will be called conidiospores.

Thallospores are asexual spores formed by transformation of existing cells of the thallus, and are set free by decay or disarticulation of the parent hyphae, not by a process of active cutting-off (abstriction). This is clearly a type of natural fragmentation of the mycelium. Thallospores may be terminal or intercalary, formed as cells of the parent hyphae. When terminal, they are not formed *de novo* from materials extruded or blown out from a sporogenous apex. When intercalary, thallospores are formed from existing cells of a hypha, singly or in chains. Since they are formed from relatively undifferentiated parts of hyphae it is to be expected that they should be widely distributed throughout the major groups of fungi. Two types of thallospores will be discussed later in connexion with the Deuteromycotina: **arthrospores** and **chlamydospores.**

Conidiospores, or 'true conidia' are asexual spores formed singly or in basipetal or acropetal succession from new elements of the thallus, i.e. from materials blown out or extruded and later abstricted from a **conidiophore.** The sporogenous part of the conidiophore is commonly apical but may be laterally placed. Several types of conidiophores and conidiospores are known, and will be discussed later under Deuteromycotina. The conidiophores may be free and discrete from one another, or aggregated to form compound sporophores called **synnemata** and **sporodochia,** or enclosed in stromatic compound sporophores called **acervuli** and **pycnidia.** Discussion of these, too, will be deferred to the later section on Deuteromycotina.

TAXONOMIC IMPLICATIONS OF ASEXUAL REPRODUCTION

Some types of asexual reproduction (e.g. fragmentation and thallospore-production) are so widely distributed over the whole range of fungi, and others (e.g. fission) so narrowly confined, that none of these has much value in establishing broad taxonomic categories; their use is mainly in distinguishing lesser taxa.

Forms of asexual reproduction by means of sporangia and sporangiospores are an excellent taxonomic character; they are confined to lower fungi, i.e. to the Myxomycota (slime-moulds), Mastigomycotina and Zygomycotina as already delimited on the characters of their thalli or hyphae, and form a striking confirmation of this grouping.

Production of conidiospores is characteristic of the higher fungi (Ascomycotina, Basidiomycotina and Deuteromycotina) as defined by hyphal characters, but is not entirely confined to these groups. Some of the lower fungi have true conidiospores, or sporangia which can function facultatively as conidiospores. Such forms are considered to be transitional and are accounted for by the hypothesis that conidiospores have evolved from sporangia, a theme that will be pursued in Chapter 9.

Taxa Characterized by Sporangia

Division Myxomycota

Among the lower fungi, the Myxomycota are recognized principally as a group of organisms in which the thallus is a non-hyphal, amoeboid plasmodium, giving rise to sporangia of various types. The sporangiospores are non-motile meiospores which liberate motile **swarmers** each with **two anterior flagella,** one shorter than the other and directed backwards. In one genus, *Ceratiomyxa*, the sporangium is monosporous, with the spore wall fused to that of the sporangium, and the whole is dispersed as a single unit.

Subdivision Mastigomycotina

The other lower fungi comprise two groups which can be distinguished by their asexual spores: the Subdivision Mastigomycotina with motile sporangiospores (**zoospores**), and the Zygomycotina with non-motile sporangiospores (**aplanospores**). These sporangiospores are borne in sporangia which may be terminal or intercalary, sessile or borne on a distinct sporangiophore. They may either burst at maturity to liberate the sporangiospores or, in some of the Mastigomycotina, may open by means of a simple exit pore or tube, or by one or more lidded pores termed **opercula** (s. **operculum**).

Zoospores

These are 'naked' in the sense that their covering is only a hyaloplasm membrane. Normally, they are uninucleate and haploid, but in one genus, *Allomyces*, which is unique in having alternation of gametothalli and sporothalli, both haploid and diploid zoospores are formed during the life-cycle.

The flagella of zoospores are fibrillar and constructed of two central and nine peripheral fibrils, each composed of subfibrils. Two general types of flagella are distinguished: the whiplash and tinsel types. The **whiplash flagellum** has a long rigid base composed of all eleven fibrils, and a short flexible end formed of the two central fibrils only. The **tinsel flagellum** has a rachis which is beset on all sides along its entire length with short hair-like subfibrils. Within the zoospore, the flagellum of each type ends in a blepharoplast which is continued into a rhizoplast, making contact with the nucleus.

The zoospores are flagellated in various ways which appear to reflect broad evolutionary groupings; at all events they form a reliable way of classifying the Mastigomycotina into three classes: Chytridiomycetes, Hyphochytridiomycetes and Oomycetes (*see* Key on p. 111).

Subdivision Zygomycotina

No motile cells occur in members of the Subdivision Zygomycotina. The asexual spores are aplanate sporangiospores or, in some instances, conidia; the sexual spores are zygospores formed by gametangial conjugation. Two classes of Zygomycotina are recognized: Zygomycetes and Trichomycetes (*see* p. 122).

Taxa Characterized by Sterile Mycelia or Conidia

Subdivision Deuteromycotina

The Deuteromycotina, or Fungi Imperfecti, is a Subdivision to accommodate sterile or conidial imperfect states in the pleomorphic life-cycles of fungi which cannot be assigned on the basis of material at hand to any other subdivision of fungi; in effect these are states of the Ascomycotina and of a smaller number of Basidiomycotina. They are recognized by having primarily septate hyphae and various kinds of thallospores or conidiospores — never sporangia which would immediately reveal them as belonging to the lower fungal taxa.

7

SEXUAL REPRODUCTION

As in so many of their other features, fungi are very varied and versatile in their methods of sexual reproduction. All these methods, however, involve the union of two compatible nuclei which may be carried in motile or non-motile gametes, in gametangia, or in somatic cells of the thallus. Three processes occur in sequence during sexual reproduction: plasmogamy, karyogamy and meiosis.

THE SEXUAL CYCLE

Plasmogamy

This is the process of anastomosis of two cells and fusion of their protoplasts to bring two haploid nuclei of opposite sex potential or mating type together in one cell.

Karyogamy

This process is the fusion of two compatible haploid nuclei into one diploid nucleus. In many lower fungi karyogamy occurs immediately after plasmogamy; but commonly, it is delayed in higher fungi, in which case the anastomosed cells each contain a dicaryon. Such dicaryotic cells may grow and divide, forming daughter cells whose conjugate pairs of nuclei replicate the original pair. Eventually, a large number of dicaryotic cells are formed and give rise to tissues. Terminal dicaryotic cells (**zeugite** cells) are formed in special layers of tissue (the hymenial layers) and karyogamy finally takes place at about the same time in all of these zeugites. The zeugite cells of Ascomycotina are called **asci** (s. **ascus**) and those of Basidiomycotina are **basidia** (s. **basidium**). Karyogamy is followed by meiosis and spore-production from the zeugites. Thus a very large number of zeugites and an even larger number of spores can originate from one original association of two nuclei during plasmogamy.

Meiosis

After karyogamy, the zygote or zeugite diploid nucleus undergoes a reduction division to form two haploid nuclei each with *n* chromosomes. Ordinary mitosis follows, or in some instances precedes, meiosis. The haploid nuclei then have walls secreted around them to form spores, or migrate to the developing spores.

TYPES OF SEXUAL ANASTOMOSIS CONCERNED IN PLASMOGAMY

Planogametic conjugation

A **planogamete** is a motile gamete, or sex cell, and planogametic conjugation (Fig. 20) is the fusion of two gametes, one or both of which may be motile. It follows that such unions are possible only when free water is present at this critical stage of the life-cycle. Fungi which have this type of plasmogamy are predominantly aquatic, occur in soil moisture films, or less often occur as parasites in the cells of higher plants where the cell sap provides the necessary moisture for the swimming gametes.

Most cases of planogametic conjugation are **isogamous,** with morphologically indistinguishable gametes. Some are **anisogamous**, with gametes which are morphologically similar but differ in size. Although **heterogamy,** involving morphologically dissimilar gametes, is by no means uncommon in fungi, **heterogamous planogametic** conjugation is unique to the genus *Monoblepharis*; this has male planogametes called **antherozoids** borne in male gametangia (**antheridia**), and female non-motile gametes called **oospheres** borne in female gametangia (**oogonia**).

The zygote resulting from isogamous or anisogamous conjugation of planogametes is a **resting sporangium,** a somewhat resistant cell which on further development functions as a sporangium by differentiating zoospores internally. The zygote resulting from heterogamous planogametic conjugation is an **oospore,** a resistant resting spore which germinates by means of a germtube and gives rise to a mycelium directly.

Gametangial contact

This type of union is also heterogamous, but neither male nor female gametes are motile. The male and female gametangia come into contact (Fig. 20) and the male gametes, consisting mainly of nuclear material, are directly transferred into the female gametangium through a pore dissolved in the common wall at the point of contact, or sometimes through a short fertilization tube. **Oogamy,** such as occurs in some of the lower fungi, is but a specialized form of gametangial contact; here the male gametes (mainly nuclear material) contained in antheridia fertilize one or more oospheres contained in oogonia. The resulting zygote cell is an oospore or a resting sporangium, depending upon the type of fungus and its environment. Gametangial contact is also a common process in some higher fungi (Ascomycotina), but here there are no oospheres and the female gametes are represented by nuclei. These remain associated in conjugate pairs with the

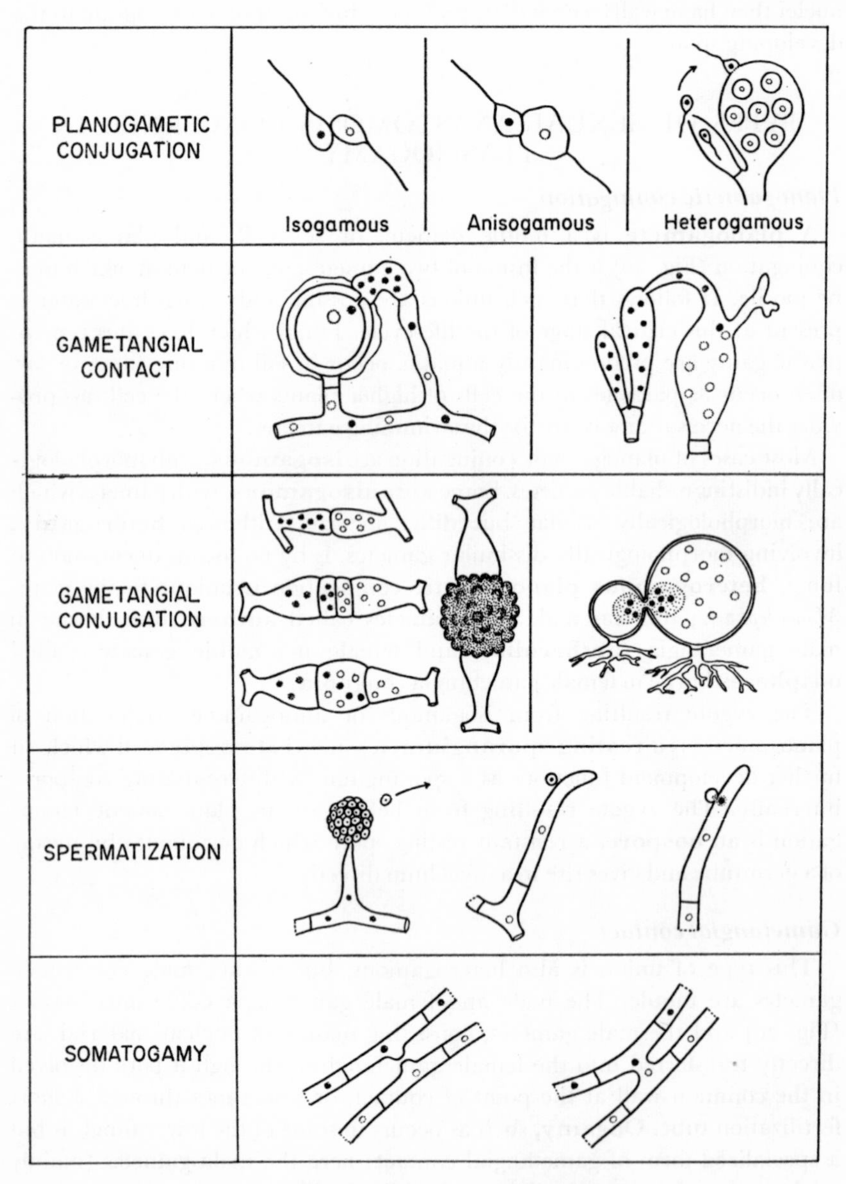

Fig. 20. Types of plasmogamy (diagrammatic).

male nuclei, dicaryotic tissues are formed, and nuclear fusion eventually takes place in the asci to form the true zygotes, diploid nuclei.

Since there are no motile gametes, gametangial contact is possible in non-aquatic fungi, but nevertheless is common in aquatic fungi and in parasites occupying cells of higher plants. The fungus thus achieves a measure of independence from free water in situations which are liable to dry up.

Gametangial conjugation

Again there are no motile gametes and this process (Fig. 20) is independent of an aquatic phase in the life-cycle. Here, the whole contents of two gametangia act as gametes and their meeting and fusion takes place in one of two ways. In the aquatic fungi (e.g. some Chytridiomycetes) the two gametangia meet and their entire contents fuse in the female gametangium, leaving the other to collapse; this results in the production of a resting sporangium as the zygote. In the Mucorales and Entomophthorales (Zygomycotina), however, which are essentially non-aquatic, the two gametangia meet and their contents pass into a common cell formed between them by dissolution of the intervening walls at the point of contact. The zygote resulting from this type of plasmogamy is a **zygospore,** a thick-walled resting spore. The zygospore may germinate directly to a mycelium or indirectly by production of a **zygosporangium** with aplanospores, each formed on a short sporangiophore.

Spermatization

Spermatization (Fig. 20) is accomplished by **spermatia** (s. **spermatium**), small haploid uninucleate male cells which are carried passively by wind, water or insects and anastomose with the wall of a female gametangium, which may be reduced to no more than a specialized **receptive hypha.** The spermatia are borne externally in various ways on **spermatiophores** and have the appearance of small conidia, but are incapable of germinating. The contents of the spermatium (mainly nuclear material) are transferred to the receptive hypha and plasmogamy results in the formation of a dicaryotic cell. The dicaryotic phase may persist for some time and is extended to other cells to form dicaryotic tissues; but eventually the true zygotes, diploid nuclei, are formed by karyogamy in special zeugite cells (asci and basidia). Spermatization is associated only with Ascomycotina and Basidiomycotina and is often independent of free water.

Somatogamy

Anastomosis of somatic hyphae is a very common phenomenon in the higher fungi (but is virtually absent from lower fungi) and it has a sexual significance only when it brings together compatible nuclei of opposite mating type into one cell. The dicaryotic cell so formed then gives rise to dicaryotic tissues and eventually to the zeugites in which karyogamy occurs and the true zygotes, diploid nuclei, are formed. Somatogamy (Fig. 20) is independent of free water and is very commonly found in the Ascomycotina and Basidiomycotina, which are essentially terrestrial. It may be regarded as a reduced but highly efficient form of sexuality.

SEX AND SEXUAL COMPATIBILITY: HETEROTHALLISM

Male and female sex organs in fungi are often clearly distinguishable on morphological grounds. The thalli may be **monoecious,** when each haploid thallus bears both male and female organs, or **dioecious,** when these organs occur on different haploid thalli. However, in perhaps the majority of fungi a sexual process takes place although the participating structures (e.g. similar gametangia, or nuclei in somatic hyphae) are not morphologically distinguishable as male or female; the thalli of such fungi can be said to be **sexually indeterminate.** To avoid assigning a sex to such thalli, one usually speaks of them as opposite mating types, or plus and minus strains.

Self-fertility in a monoecious thallus has the disadvantage of limiting the number of new recombinations of alleles during sexual fusion and meiosis, so that there is likely to be less genetic variation in the progeny than if nuclei from different thalli had fused. Genetic variation is often an advantage to the organism since it is the basis of evolution by natural selection. Against this, however, may be set the greater likelihood of fertilization in a monoecious organism and the fact that genetic stability may be an advantage to an organism which occupies a special, stable environment.

In many fungi, the disadvantages of self-fertility have been overcome by their being self-sterile. Self-sterility may be achieved merely because the thalli are dioecious, or because the opposite mating types occur as different thalli; but alternatively, or in addition, self-sterility may occur in monoecious, dioecious or sexually indeterminate thalli in which inbreeding is controlled by one or more genetic factors for compatibility. The opposite strains or sexes may well be present but incompatible, so sexual fusion cannot take place successfully. On the basis of compatibility we can group fungi as either homothallic or heterothallic, as follows: **homothallic** species are those with only one strain of thallus as regards compatibility type, and this is both self-fertile and self-compatible; **heterothallic** species are those with more than one strain of thallus, whether the difference in strains is one of sex (dioecism or morphological heterothallism), compatibility (haploid incompatibility or physiological heterothallism), or both. Every strain of a heterothallic species is self-sterile, requiring the aid of another thallus of opposite sex or mating type, opposite compatibility factors, or both, for sexual fusion to occur.

Heterothallism can lead to curious situations when strains of a single species have become geographically separated. An example is that of *Fomes pinicola* (Mounce and Macrae, 1938), which exists as two intersterile strains in North America, each of which is interfertile with a third strain that occurs in Europe. This example shows how difficult it may sometimes be to determine species which are based on the criterion of interfertility.

Segregation of compatibility factors: polarity

When only one pair of alleles, *A* and *a*, controlling compatibility occur at a single chromosome locus, the species is said to exhibit **bipolar heterothallism** (Fig. 21). The appropriate chromosome of the one strain of thallus carries the allele *A*, and that of the other thallus carries the allele *a*; only thalli carrying complementary alleles of this pair are compatible and interfertile, giving an *Aa* zygote. There is a 50% chance of interfertility between opposed mycelia in the species, i.e. only half of the encounters can be successful. This form of heterothallism was originally discovered by Blakeslee (1904) in the Mucorales.

Tetrapolar heterothallism (Fig. 21) involves two compatibility loci each with a pair of alleles, *A* and *a* or *B* and *b*, controlling compatibility. Therefore, the species may exist as four strains of thalli with the respective allelic constitutions *AB*, *Ab*, *aB* and *ab*; only those thalli which combine to give an *AaBb* zygote are compatible and interfertile. There is a 25% chance of interfertility between opposed mycelia of the species, i.e. only a quarter of the encounters are successful. Both bipolar and tetrapolar heterothallism have been commonly encountered in members of the higher fungi, particularly the Basidiomycotina.

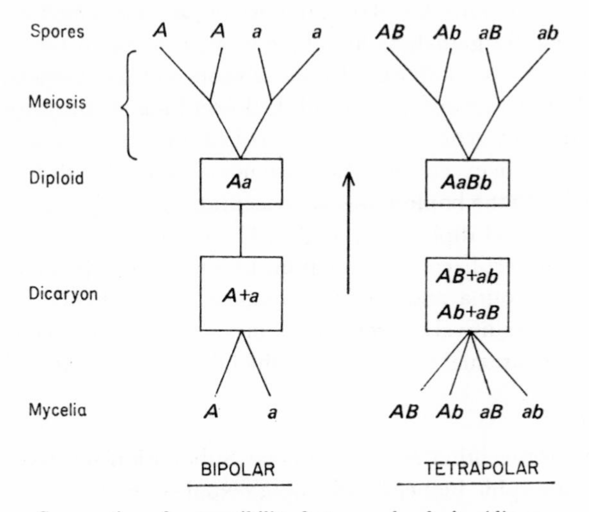

Fig. 21. Segregation of compatibility factors and polarity (diagrammatic).

In the higher fungi a further complication is introduced by the fact that with both bipolar and tetrapolar heterothallism a great number of alleles (not simply one or two pairs) may occur at one or two compatibility loci respectively — a phenomenon responsible for **multiple allelomorph heterothallism.** So many alleles may be concerned that there is a very great increase in the chance of interfertility between opposed strains, but a reduced chance of interfertility between strains arising from spores derived from the same thallus.

The outbreeding mechanisms of fungi are therefore seen to be first, dioecism and second, heterothallism of various kinds. Garrett (1963) neatly summarizes heterothallism by saying: 'Heterothallism promotes the occurrence of outbreeding, and therefore subserves the same end as the sexual process, which it renders more efficient. Heterothallism is not the same as sex; it is a refinement superimposed upon it.'

TAXONOMIC IMPLICATIONS OF SEXUAL REPRODUCTION

Although characters of sexual reproduction are generally considered to be the most definitive for taxonomic purposes, it should be remembered that the major groups of fungi are also recognizable by characters of their hyphae or thalli, and their asexual reproduction.

Lower Fungi

In the Myxomycota, planogametic conjugation of either swarmers or myxamoebae results in a motile zygote. In the Mastigomycotina, the methods of plasmogamy are planogametic conjugation, gametangial contact, and less frequently gametangial conjugation, leading to the production of resting sporangia or oospores. In the Zygomycotina, gametangial conjugation results in a zygospore. None of the lower fungi exhibits spermatization or somatogamy. In general one act of fertilization produces one sexual spore and, with very rare exceptions, these are not associated with fruitbodies. The thallus belongs to the haplophase (an exception is in *Allomyces*, with alternating haplophase and diplophase thalli). The diplophase is usually brief and confined to the fertilized gametangium or the sexual spore before meiosis. The Mastigomycotina and Zygomycotina are, respectively, predominantly oogamous and exclusively zygogamous, but are even better characterized by their methods of asexual reproduction which have already been described.

Higher Fungi

The Deuteromycotina are an artificial Subdivision set apart for asexual states of pleomorphic higher fungi whose sexual states are either unknown or not commonly found.

Those higher fungi which do produce sexual states can be divided into two groups according to their types of zeugites and their manner of forming 'sexual' spores. These are the Ascomycotina and Basidiomycotina.

In some Ascomycotina, plasmogamy takes place by gametangial conjugation (e.g. in yeasts); in others, gametangial contact, spermatization, or somatogamy may occur. In Basidiomycotina, there are never sex organs or motile cells, and only spermatization or somatogamy are possible; both are processes by which dicaryotization of the hyphae takes place. Gametangial contact can also result in dicaryotization of part of the thallus. In both Ascomycotina and Basidiomycotina, the dicaryotic phase can persist for a long time and the individual may then consist of partly monocaryotic, partly dicaryotic and sometimes partly heterocaryotic multinucleate cells, depending on the circumstances. Many variations in the nuclear condition of hyphae are possible. Fruitbodies are usually produced and consist mostly of aggregated dicaryotic hyphae resulting from further development of the original dicaryotized cells formed by plasmogamy, but in addition there may be supporting monocaryotic or heterocaryotic hyphae in the fruitbodies. The diploid is confined to the fusion nuclei produced by karyogamy in the terminal zeugites of the fruitbody. These terminal cells (Fig. 22) are known as **ascus mother cells** in the Ascomycotina and **probasidia** in the Basidiomycotina. They are usually formed in large numbers at about the same time and the same level in the fruitbody, forming the stratum known as the **hymenium.** It will be observed that it is possible for one dicaryotic cell, arising from one act of nuclear association by plasmogamy, to give rise to innumerable zeugites and sexual spores.

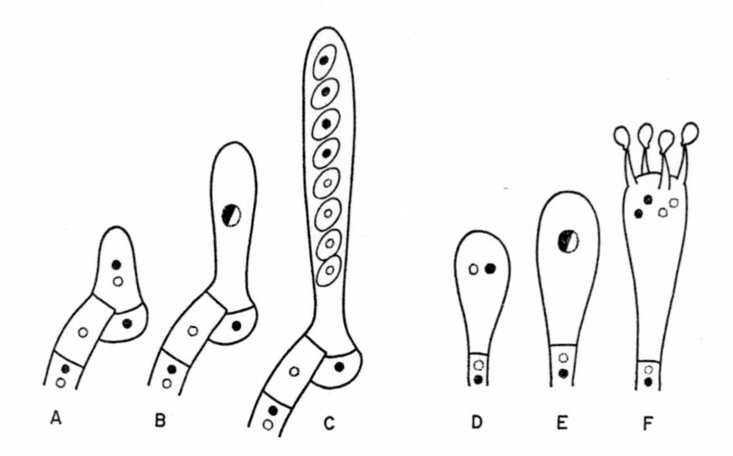

Fig. 22. Formation of asci (A, B, C) and basidia (D, E, F). A, dicaryotic zeugite; B, diploid ascus mother cell; C, ascus with ascospores; D, dicaryotic zeugite; E, diploid probasidium; F, basidium (metabasidium) with four basidiospores on sterigmata.

The ascus mother cell enlarges into an **ascus** (p. **asci**) in which the diploid nucleus divides by meiosis and mitosis to form a number of haploid nuclei (usually eight or fewer) which have walls secreted about them and develop into **ascospores** contained within the ascus. The probasidium enlarges into a **basidium** (p. **basidia**) in which a small number of haploid nuclei (usually four, sometimes fewer, or as many as eight) are similarly formed by meiosis and mitosis. These then migrate to the developing **basidiospores** which are formed externally at the ends of small tubes, or **sterigmata** (s. **sterigma**), which are outgrowths of the basidial wall.

Nuclear behaviour is not always clear-cut and according to the most common patterns outlined above. For instance, in some Ascomycotina the number of spores per ascus may be a large power of 2 perhaps as many as 2^8, or nuclei may abort and the result will be fewer spores than is usual for the species. More than one nucleus may occasionally be included in a single spore, and this will affect its genetic constitution and that of any thallus derived from the spore. Heterocaryons arise in various ways and may change their nuclear constitutions from time to time. All these factors have wide implications in causing genetic variation among members of a species.

One of the hallmarks of the higher fungi is the production, in most cases, of fruitbodies; and within each of the two Subdivisions Ascomycotina and Basidiomycotina much of the classification rests upon distinctions between different types of fruitbody.

8

FUNGI WITH A PLASMODIAL THALLUS DIVISION MYXOMYCOTA: SLIME-MOULDS

Affinity with other organisms

Two distinct phases occur in the life-cycle of a slime-mould. First there is an active assimilative phase in which the thallus, a protozoa-like amoeboid **plasmodium,** moves with a creeping motion and feeds either saprobically or holozoically by engulfing and digesting bacteria or other solid particles in food vacuoles formed according to need. Second, there is a quiescent phase in which the protoplasm of the plasmodium becomes concentrated into fungus-like sporangia with dry sporangiospores which are meiosporous in their formation. Hyphae are not formed at any time in the life-cycle. Because they appear to have affinities with protozoa and with fungi, the slime-moulds have always been peculiarly difficult to classify. Further difficulties arise because some slime-moulds (Class Myxomycetes) are free-living while others (Class Plasmodiophoromycetes) are endoparasites in cells of higher plants and also differ in several other respects from the free-living types. In addition, two other Orders of slime-organisms, the Acrasiales and the Labyrinthulales, are often considered in this context.

Martin (1932, 1960) has pointed out that if the algal origin of fungi is accepted, then the slime-moulds are not fungi; but if the fungi have been descended from colourless flagellates, the Myxomycetes can well be regarded as fungi, derived from the same general ancestral types as other fungi, but independent and highly specialized. The plasmodium differs from the young coenocytic assimilative phase of most Mastigomycotina and Zygomycotina mainly in the lack of a true cell wall; on the other hand some of the Chytridiomycetes have naked thalli with amoeboid movement. Enzymes can be used to dissolve the walls of hyphae of mycelial fungi, leaving an amoeboid mass of living protoplasm which can later regenerate a wall. One can postulate a series of organisms in which the hyaloplasm membrane enclosing a plasmodium became thickened and differentiated as a true wall surrounding a coenocytic cell. This, if true, does not imply that other fungi have been derived directly from Myxomycetes, but merely that there is a possibility that both arose from similar ancestral stock. The absence of a cell wall is not a good reason for excluding the Myxomycota from the category of fungi, and its absence has allowed these organisms to live, at least at times, in a holozoic manner.

MYXOMYCETES

Typical situations in which Myxomycetes occur are damp soil or especially damp rotten wood, humus and compost, where their plasmodia feed on bacteria which are partly responsible for decaying the vegetable matter. They are widely distributed, many species being cosmopolitan. They appear to have no direct economic importance but this may merely be because their microbiological activities have never been adequately investigated. Because their plasmodia are a source of large masses of relatively pure protoplasm, they have been used in several fields of biological research. Occasionally, gardeners are perturbed to find the large purple-brown aethalia of *Fuligo septica* (Fig. 27) in vegetable beds, the iridescent sporangia of *Diachea leucopoda* (Fig. 24B) on strawberry leaves, or the grey sporangia of *Physarum cinereum* (Fig. 24A) disfiguring large areas of lawn. But all these are harmless and disappear of their own accord if watering is reduced, or if lawns are mown. Yet people are often reluctant to accept that Myxomycetes are harmless, or that they may suddenly appear after a spell of warm, wet weather. In January, 1969, a householder in Adelaide, South Australia, heard strange noises during the night and was alarmed when he found next morning that there were patches of 'sludge' on his lawn which later dried to a greyish powdery deposit. He immediately concluded that this was exhaust sludge deposited by a 'flying saucer'; most significant, the deposit was found to contain carbon when analysed for him by a chemist. He, and several others who had a similar experience, are still not convinced that the sludge was the plasmodium of *Physarum cinereum*, and the grey powder its sporangia.

Typical Structure and Life-Cycle

The Myxomycete fructification produces dry sporangiospores which are probably wind-distributed. In films of water a spore may germinate by setting free either a non-flagellated amoeboid cell called a **myxamoeba** (p. **myxamoebae**), or one or more **swarmers** each with **two anterior flagella,** one shorter than the other and directed backwards (Fig. 23). A myxamoeba may put out flagella and become a swarmer, or a swarmer may retract its flagella and become a myxamoeba. Both types of cell have a hyaloplasm membrane, and both may become encysted under unfavourable

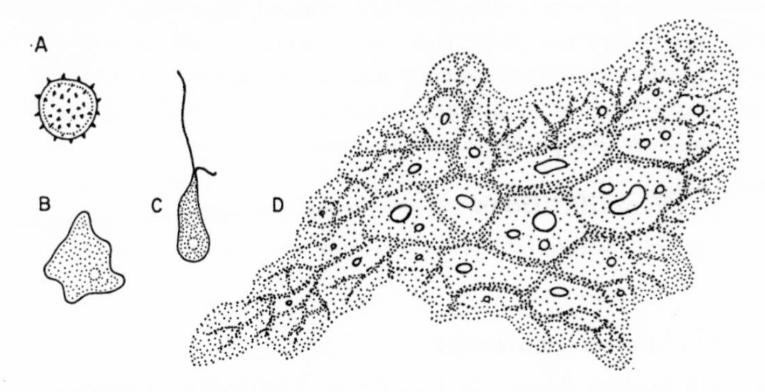

Fig. 23. Stages in life-cycle of a Myxomycete (diagrammatic). A, sporangiospore; B, myxamoeba; C, swarmer with two anterior flagella; D, plasmodium with vacuoles and 'veins', and a pseudopodium at lower left.

conditions. Either when flagellated or amoeboid these haploid cells may act as gametes and fuse in pairs of their respective types to produce a diploid cell as zygote (planogametic conjugation). Repeated mitosis of the diploid nucleus occurs and the mass of multinucleate protoplasm grows and expands into a diploid plasmodium (Fig. 23) covered by a hyaloplasm membrane.

Plasmodium

The plasmodium is amoeboid in the sense that it moves by extending pseudopodia, lacks a true cell wall, and feeds holozoically at least part of the time. It may enlarge either by coalescence of several diploid zygotes (thus becoming a syncyte), or by growth of a single zygote whose nucleus divides repeatedly by mitosis (thus a coenocyte); which process occurs is probably a matter of circumstances or of species.

Three types of plasmodia have been noted by Alexopoulos (1962). The simplest is the **protoplasmodium** which is microscopic and homogeneous, produces a single sporangium, and lacks 'veins'. The **aphanoplasmodium** is small at first but later enlarges into a network of fine strands; the 'veins' are present but inconspicuous, as is the whole plasmodium, and the protoplasm shows reversible streaming movements. The largest and most spectacular type is the **phaneroplasmodium.** This is often several centimetres in diameter, a slimy, viscid, often brightly coloured mass. Its protoplasm is distinct and highly granular, and is composed of parts of differing viscosity. The less dense parts form a system of branching channels ('veins') bounded by denser parts, forming a network with the larger meshes towards the centre of the plasmodium and smaller and more numerous ones towards the periphery. Within the 'veins', protoplasm and granular substances flow rhythmically to and fro. In porous substrata, such as rotten wood, the plasmodium may flow through the pores, now diverging and then coalescing again into one large mass, eventually coming to the surface and becoming quiescent just before the sporangia are formed. Under unfavourable conditions the plasmodium may contract into a hard mass, a 'sclerotium', which becomes restored again to the active plasmodial state under suitable conditions of moisture, temperature and nutrition.

Sporangiate Fructifications

Four types of sporangiate fructifications are known among the Myxomycetes:

Simple multisporous sporangia

These are formed when the plasmodium becomes raised into a large number of crowded but discrete sporangial primordia (Fig. 24). In stipitate forms, the protoplasm flows upwards from the primordium and deposits a central shaft of non-living material which forms the sporangiophore. After the bulk of the protoplasm has reached the apex an outside wall, or **peridium,** is secreted around the developing sporangium. The peridium is a delicate non-cellular membrane, often with lime or other chemical inclusions. The protoplasm within the peridium becomes vacuolated so that a series of cleavage furrows is formed. At about this stage the nuclei undergo meiosis and each haploid nucleus is finally isolated in a small block of protoplasm separated by furrows from the others. Finally, walls are secreted around the blocks of protoplasm to form sporangiospores. These are usually coloured and may be either smooth or ornamented. In most Myxomycetes refuse material is deposited in the larger cleavage furrows and hardens into a system of non-living threads known as the **capillitium,** which although hypha-like, are not hyphae. The type of capillitium is an important diagnostic feature since the threads may form distinctive types of networks which are constant for the species or genera. They may have swellings (**lime-knots**) in which lime is deposited, and may be attached in different ways to the peridium or to the **columella,** a prolongation of the sporangiophore within the sporangium. The capillitium may otherwise be present in the form of free threads known as **elaters.** Capillitial threads, and especially elaters, may be characteristically ornamented with spiral bands or spines, or with both. The capillitium is generally springy and sometimes hygroscopic; both these features assist in spore-dispersal once the delicate peridium is ruptured. Not only does the capillitium liberate spores by its movements, it also retains some of the spores in its meshes to be liberated at later intervals. Presumably, the species has a greater chance of survival if all the spores are not liberated at once. In many species the remnants of the plasmodium form a thin, translucent layer known as the **hypothallus** at the base of the sporangiophores.

Not all sporangia of this type are stipitate; some are sessile and in some

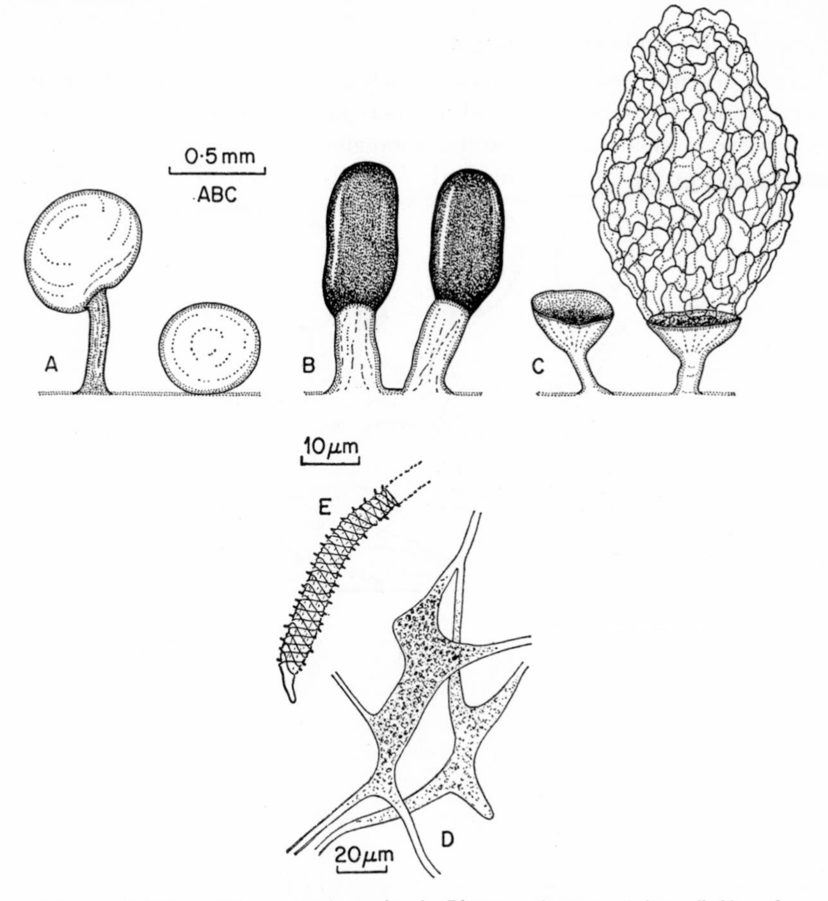

Fig. 24. Simple multisporous sporangia. A, *Physarum cinereum*, stipitate (left) and sessile sporangia; B, *Diachea leucopoda*; C, *Arcyria* sp., stipe and cupular base of empty sporangium (left) and the same with expanded capillitium (right); D, *Physarum cinereum*, lime-knots in capillitium; E, an elater of *Trichia* sp.

species approach the type of sporangiate structure known as a plasmodiocarp (described below); indeed sessile or stipitate sporangia, and plasmodiocarps, may be present sometimes in a single collection. All, however, are usually small and more or less symmetrical in such cases.

Some important points to note are: the capillitial threads are composed of waste chemical material and are not hyphae (compare capillitium of Gasteromycetes); the sporangiospores are meiospores (compare sporangiospores of Mastigomycotina and Zygomycotina); and the sporangiospores are not produced from the capillitial threads.

Simple monosporous sporangia

In the genus *Ceratiomyxa*, only, a delicate branching sporophore arises erectly from the substratum and at intervals bears small denticles, each of which has a single monosporous sporangium at its apex (Fig. 25). The sporophore appears to be composed of non-living material and is thought to

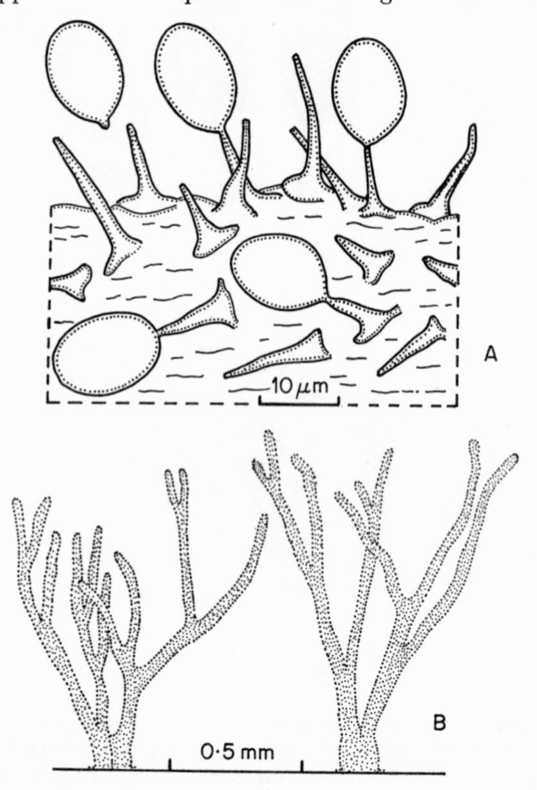

Fig. 25. Simple monosporous sporangia of *Ceratiomyxa fruticulosa*. A, sporangia arising from denticles on the branched hypothallus; B, habit of branched hypothallus.

represent an erect, branched hypothallus. The sporangia used to be regarded as exogenously formed spores, but the studies of Gilbert (1935) make it almost certain that they are sporangia, each containing a single endogenous spore whose wall is fused to that of the sporangium itself.

Plasmodiocarps

In forming a plasmodiocarp (Fig. 26) the protoplasm of the plasmodium becomes concentrated around some of the main 'veins' and is then covered by a peridium. A plasmodiocarp is therefore sessile, but differs from an ordinary sessile sporangium in being asymmetrical, linearly extended and in

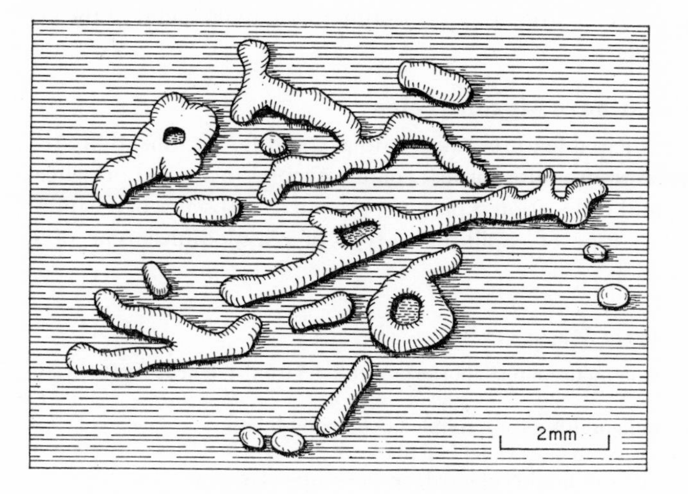

Fig. 26. Plasmodiocarps of an unidentified Myxomycete.

retaining to a certain degree some of the branching habit of the plasmodial 'veins' so that it is branched or reticulate. The internal structure is like that of a sporangium. Sessile sporangia and short plasmodiocarps showing little branching may sometimes be found in a single collection; in such cases the plasmodiocarps are distinguished by being more elongate than the sporangia and usually curved, Y-shaped or doughnut-shaped. *Hemitrichia* is one genus in which plasmodiocarps are well shown.

Aethalia

When the entire plasmodium masses together into one or only a few primordia, aethalia (s. **aethalium**; Fig. 27) are formed. These are sessile, large, globose or hemispherical structures, representing a type of compound sporangium covered by a common, firm or fragile peridium. Inside some aethalia the walls of confluent sporangia may be evident; in others they may

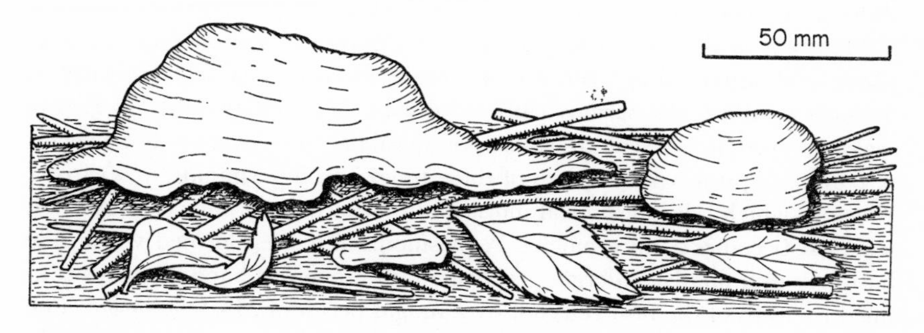

Fig. 27. Two aethalia of *Fuligo septica*.

be degenerate or absent. Good examples are to be found in the genera *Fuligo* and *Lycogala*.

Classification of Myxomycetes

The principal features taken into account for classification are: sporangia monosporous or multisporous; colour of spores in a mass; presence or absence of lime, and its location if present; presence or absence of a capillitium and a columella; type of capillitium; type of plasmodium; type of sporangiate fructification.

Key to orders of Myxomycetes

1.	Sporangia monosporous, borne on denticles from a delicate branched hypothallus	*Ceratiomyxales*
	Sporangia multisporous, simple or compound	2.
2.	Spores rusty to deep violet in a mass	3.
	Spores pale or bright coloured, not rusty or violaceous	4.
3.	Without lime, except possibly in the sporangiophore	*Stemonitales*
	With lime in the peridium or capillitium	*Physarales*
4.	Columella and capillitium absent	*Liceales*
	Columella absent, capillitium present	*Trichiales*
	Columella usually present, capillitium present, sporangia very small	*Echinosteliales*

PLASMODIOPHOROMYCETES

The Plasmodiophoromycetes, or endoparasitic slime-moulds, form a small group whose affinities with other organisms are uncertain largely because details of their life-cycles are still a matter of controversy. We have chosen to include them in the Myxomycota because they have plasmodia and because their swarmers (which may also be able to act as gametes) each have a pair of anterior whiplash flagella of unequal size. In contrast to the Myxomycetes, however, sporangiate fructifications are lacking, the spores being confined within the host cell wall; it is also possible that haplophase and diplophase plasmodia alternate in the life-cycle; finally, it is probable that the spore walls differ in composition from those of Myxomycetes.

Members of the Plasmodiophoromycetes parasitize various algae, marsh plants and aquatic fungi, but are of especial interest because *Plasmodiophora brassicae* produces the serious clubroot disease of crucifers, and *Spongospora subterranea* produces a scab disease of potato tubers. *Spongospora* is also a vector of potato mop top virus. Spores of these fungi in soil germinate to set free a swarmer which migrates in soil moisture films and finally infects a host root where it becomes a myxamoeba. A plasmodium is then formed within the host cells and finally undergoes cleavage into numerous haploid spores confined by walls of the host cell, not by a fungal peridium. In *Spongospora* (Fig. 28) the spores are aggregated into a few large spore-balls within the host cell,

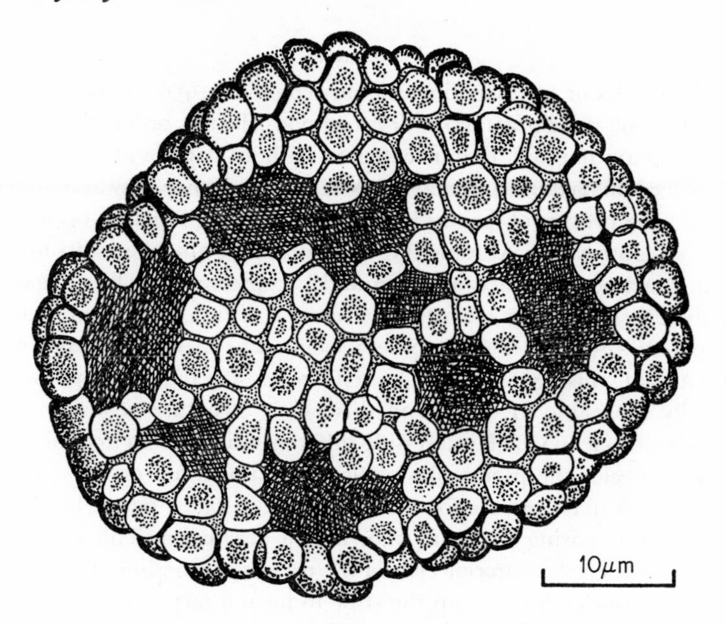

Fig. 28. *Spongospora subterranea*, a single spore-ball taken from a sorus occupying a host cell.

each appearing like a hollow sphere composed of spores. In *Plasmodiophora* the spores remain discrete.

COLONIAL SLIME-MOULDS: LABYRINTHULALES AND ACRASIALES

These two orders of organisms are often studied in connexion with the Myxomycota and have been more or less neglected except by mycologists. Their slimy consistency in some stages of the life-cycles is suggestive of a plasmodium, but otherwise they have little in common with Myxomycota or other fungi and are probably better classed with amoeboid Protozoa.

Labyrinthulales

These net slime-moulds are mostly parasites of marine plants. Their somatic cells are spindle-shaped in almost all species, and are naked and uninucleate. From each end of these cells long slimy filaments are put out and join together to form a slimy network. The spindle cells glide along the filaments but in effect they form a colony of cells held together by the network, which is sometimes called a **net-plasmodium;** no true plasmodium exists, however. At length the spindle cells become densely aggregated and encapsulate to form spores within a sorus. No sexual process is known.

Acrasiales

The Acrasiales or cellular slime-moulds (Bonner, 1959) occur as saprobes in soil, litter, on dead grass or on dry seed pods. In this order, the amoeboid somatic cells are haploid, uninucleate and without flagella at all stages. The amoebae feed on bacteria and may become encysted; they may also reproduce asexually by a process of fission. Whether sexual reproduction occurs is still a matter of controversy. At some stage certain of the amoebae start to secrete a chemical attractant called **acrasin** and become centres of aggregation to which the other amoebae are attracted along a diffusion gradient of acrasin. The grouped amoebae then become associated as a **pseudoplasmodium** without actually fusing together. The cells composing the pseudoplasmodium are readily separable in water and this whole structure is simply a slimy colony of amoebae in which interesting differentiation of function takes place.

In some species the pseudoplasmodium remains more or less stationary, but in others it may migrate for a considerable distance before being differentiated into a sporing structure known as a **sorocarp.** This is a stipitate structure in which the anterior cells of the pseudoplasmodium form the stipe and the posterior cells move up the stipe to form a terminal head of spores. The stipes may be unbranched, or may branch in various ways, with a head of spores at the apex of each branch. In most species the stipes are cellulosic and composed of large vacuolated cells, but in a few they are non-cellular. There may be a disc at the base of the stipe in some species. The heads of spores are known as **sori** (s. **sorus**). The spores have smooth cellulosic walls; on germination each gives rise to a single amoeba.

The Acrasiales have been the subject of many interesting experiments relating particularly to morphogenesis. It was possible, for example, to determine which cells of the pseudoplasmodium give rise to the stipe, and which to the sorus, by means of a very elegant experiment. Some pseudoplasmodia of *Dictyostelium* were coloured red by being fed on red bacteria, while others were left colourless by being fed on white bacteria. In the migration stage, the pseudoplasmodia were cut in half and a red anterior half was grafted to a white posterior half, and *vice versa.* In this way the flow of cells could be accurately traced during the course of differentiation of the sorocarp into stipe and sorus.

Composite pseudoplasmodia have been produced by blending different species together in a mortar, but such a composite gives rise to the sporing structures of only one of its components.

9

FUNGI WITH SPORANGIA: EUMYCOTA SUBDIVISIONS MASTIGOMYCOTINA AND ZYGOMYCOTINA

Members of these Subdivisions, together formerly known as 'Phycomycetes', include what are usually thought to be the most primitive types of Eumycota, even though the simplest show much diversity of form and complexity of life-cycle.

The chief features distinguishing these Subdivisions from other Eumycota have been mentioned in previous chapters and can be summarized thus: the thallus may be unicellular or mycelial but its cells are almost always multinucleate and septation is predominantly adventitious; asexual reproduction is by means of motile or non-motile sporangiospores; sexual reproduction results in the formation of resting sporangia or resting spores (oospores or zygospores); in both Subdivisions examples can be found where sporangia are able to act facultatively as conidia.

The two Subdivisions can be distinguished from each other as follows:

(*a*) Sporangia with zoospores; predominantly oogamous but exhibiting all types of plasmogamy except spermatization and somatogamy. *Mastigomycotina*

(*b*) Sporangia with aplanospores; entirely zygogamous, plasmogamy only by gametangial conjugation. *Zygomycotina*

As one reviews examples of fungi placed in these Subdivisions certain broad evolutionary trends become apparent. The morphology and method of reproduction are well correlated with the habitat. Evolution has apparently proceeded from simple unicellular holocarpic forms, some even without a cell wall at least in somatic stages, to complex eucarpic forms in which absorbing, anchoring, somatic and reproductive parts are well differentiated. Such change in form has evolved in parallel with evolution from an aquatic habitat to complete adaptation to land life as saprobes (most Mucorales) or parasites in plants (Peronosporales) or animals (Entomophthorales). The changes have involved considerable modification of the life-cycles, especially a change from motile zoospores to non-motile sporangiospores, or to the sporangium itself becoming deciduous, failing to differentiate sporangiospores and instead germinating in the manner of a conidium. Similarly, there has been much change in the methods of sexual reproduction and in the organs associated with it. Reproductive methods involving motile gametes have given place to methods such as oogamy or zygogamy, where plasmogamy takes place between non-motile structures which are often immersed in host tissue and where the zygotes are resistant spores well able to survive moderate drought, heat or cold.

Gametangial contact and gametangial conjugation offer several advantages over planogametic conjugation: the gametes find each other more surely and the sexual spores, though fewer, are better formed and better protected. These factors have facilitated evolution from water to other habitats, from water dispersal to wind dispersal, and from a saprobic to a parasitic existence. Despite these biological advantages, some mycologists regard this evolutionary series as showing sexual degeneration, indicated by the whole spore mother cell (gametangium) taking the function of the part (a single gamete). Also, in both oogamous and zygogamous series, karyogamy and meiosis may be retarded and shifted in position into outgrowths of the zygotes in the form of sporangia or vesicles. It is of interest to note that in the Ascomycotina and Basidiomycotina karyogamy and meiosis are still further retarded and shifted in position, and decline of overt sexuality is accompanied by increased complexity of the fruitbodies, which are largely built up of tissue formed from dicaryotic hyphae after plasmogamy. These changes are, nevertheless, apparently highly successful.

In the following pages I shall use examples to review some of these interrelated changes in form, function and habitat, rather than consider a number of illustrative life-cycles in detail.

SUBDIVISION MASTIGOMYCOTINA

Members of this Subdivision are distinguished by having motile reproductive cells in the asexual cycle, and sometimes in the sexual cycle. Dependence upon free water in at least the asexual cycle has to some extent limited the type of environment in which these fungi are able to live. Some notable evolutionary changes are associated with methods of plasmogamy, which eliminate the need for motile gametes and dependence on the presence of free water in the sexual phase of the life-cycle.

The simplest forms of Mastigomycotina are unicellular and holocarpic; more complex forms develop a system of rhizoids or a rhizomycelium, and finally eucarpic mycelial forms predominate. Some are entirely aquatic as saprobes, or as parasites of algae or fish. Others inhabit soil moisture films and commonly cause root-rots or damping-off diseases of seedlings. Some are still further adapted to land life as saprobes, or as parasites of higher plants where again the thallus is often much reduced in size and cell sap provides the required fluid for any motile reproductive cells. In some endoparasitic types the thallus may be holocarpic, unicellular or even devoid of a cell wall.

There is more variety in the methods of sexual reproduction in the Mastigomycotina than in other groups of fungi. Plasmogamy may occur by isogamous, anisogamous or heterogamous planogametic conjugation, by gametangial conjugation or by gametangial contact, resulting in a resting sporangium or an oospore zygote. Often there is physiological control of sexual reproduction: change in temperature, lessening of water supply and depletion of food may control its onset. With parasitic forms the host plant is often an annual which dies towards winter, and the sexual spores are thick-walled resting spores (oospores) which are well suited to over-wintering; the asexual cycle predominates in the summer. In Mediterranean climates, however, where there is a mild, wet winter and a hot, dry summer, the position is often reversed and oospores, if produced at all, are the over-summering spores.

Thallospores are sometimes formed, but typically, the Mastigomycotina reproduce asexually by means of zoospores in sporangia. The sporangia may be terminal or intercalary, sessile or borne on a sporangiophore, and may either burst irregularly to liberate the zoospores at maturity or may open by means of simple pores or operculate pores.

Three classes of Mastigomycotina can be distinguished as follows:

(*a*) *Chytridiomycetes:* Fungi with posteriorly uniflagellate (whiplash type) zoospores and similarly flagellated planogametes. This grouping is correlated with sexual reproduction by planogametic or gametangial conjugation, resulting in a resting sporangium or an oospore.

(*b*) *Hyphochytridiomycetes:* Fungi with anteriorly uniflagellate (tinsel type) zoospores. Sexual reproduction not known. Examples: *Hyphochytrium*; *Rhizidiomyces*. Not considered further.

(*c*) *Oomycetes:* Fungi with biflagellate zoospores, each with one tinsel and one whiplash flagellum. In some of the more advanced members the sporangia are deciduous and dispersed as a whole. According to circumstances these may either become differentiated internally into zoospores or may remain undifferentiated within, and thus function as a single conidium germinating by a germtube. This type of sporangium with a dual function is called a **conidiosporangium**. In some other series of Oomycetes the swarming period of the zoospores is progressively reduced until finally the zoospores become encysted within the sporangium itself and lack any motile phase; these may then be dispersed as aplanate sporangiospores. Sexual reproduction in the Oomycetes is almost always heterogamous by means of gametangial contact, and results in the formation of an oospore.

CLASS CHYTRIDIOMYCETES

All 'chytrids' have planogametes and posteriorly uniflagellate zoospores, and depend upon free water during the reproductive phases. Many are entirely aquatic, while others inhabit soil moisture films or are parasitic on or in algae or land plants. There is great variety in the methods of sexual reproduction.

Order Chytridiales

In the simplest **monocentric** chytrids the thallus is a single holocarpic cell, which may even be amoeboid and lack a true cell wall in some endoparasitic forms. More advanced types of monocentric chytrids (e.g. *Rhizophydium*; Fig. 29) are eucarpic with a thallus composed of a single cyst-like cell and rhizoids, while in the **polycentric** chytrids the thallus is composed of several such cells connected by a system of rhizomycelia. A true mycelium is lacking in all members of the Chytridiales.

Olpidium brassicae (Fig. 30) is important as a cause of root disease in crucifers and other plants and as a vector of lettuce big vein virus, tobacco necrosis virus and tobacco stunt virus. *Synchytrium endobioticum* causes the serious potato wart disease and is a vector of potato virus X. In both these species, zoospores infect the host plant and develop into unicellular thalli within the host cells. These thalli become converted as a whole into sporangia with zoospores, or gametangia with isogamous planogametes. In *Olpidium* there is a single sporangium which liberates the zoospores through an exit tube to the exterior of the host, while in *Synchytrium* the sporangia are grouped in a **sorus** and liberate zoospores by rupture of their walls. On liberation, the zoospores encyst for a time before germinating and infecting a new host. The onset of the sexual reproductive phase is controlled by the environment, taking place

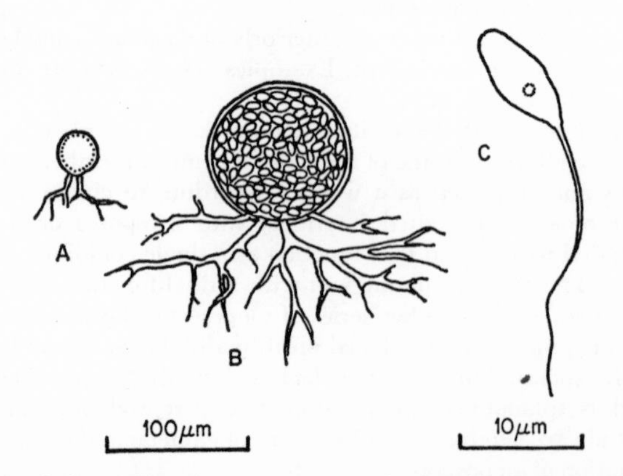

Fig. 29. *Rhizophydium sphaerotheca*, a monocentric chytrid. A, B, small and large sporangia with rhizoids; C, zoospore with one posterior whiplash flagellum.

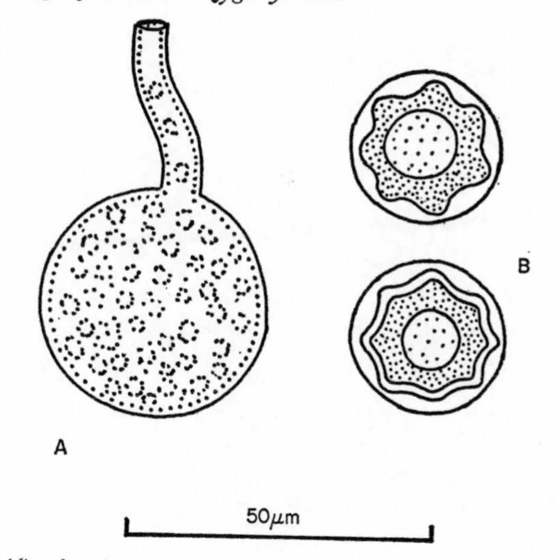

Fig. 30. *Olpidium brassicae*. A, sporangium with exit tube; B, two resting sporangia.

under relatively dry conditions. The isoplanogametes conjugate in pairs to form a motile zygote with two flagella. This rests upon the host, withdraws its flagella, and its contents enter the host and become converted to resting sporangia which give rise to zoospores on germination. Resting sporangia of *Synchytrium endobioticum* have been found to be viable after as long as 10 years in soil; the implications of this from the practical aspect of control, are obvious.

Species of *Rhizophydium* (Fig. 29) have a thallus composed of a cyst-cell and rhizoids. In the asexual phase the cyst-cell acts as a sporangium with a variable number of papillae in the wall through which the zoospores are liberated. The zoospores grow into new thalli or, in the sexual phase, into gametangia which fuse by gametangial conjugation to form a resting sporangium as the zygote. After meiosis the resting sporangium liberates zoospores.

Species of *Physoderma* are important pathogens of lucerne, clover and maize. Figure 31 shows a sorus of resting sporangia of *Physoderma alfalfae* in tissue of lucerne.

Order Blastocladiales

Most of the Blastocladiales are inhabitants of water or soil. The genus *Allomyces* is of special interest because plasmogamy takes place by conjugation of anisogamous planogametes, and there is a unique alternation of gametothalli and sporothalli. Diploid and haploid zoospores are also produced. Both types of thallus consist of an erect stem arising from rhizoids and branching dichotomously into branches which bear the reproductive organs. The

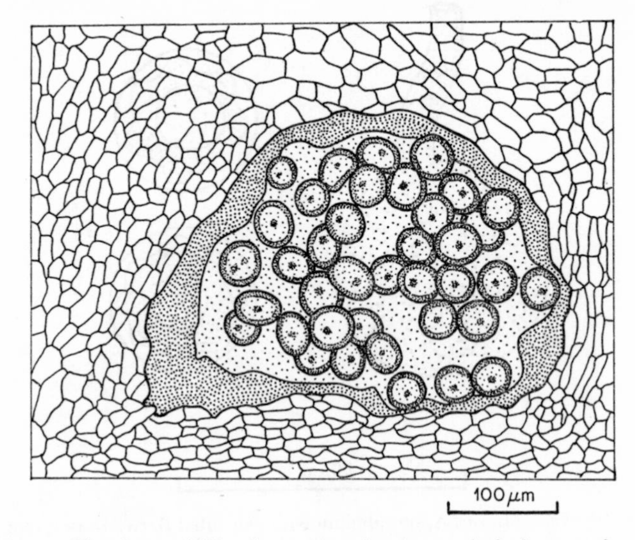

Fig. 31. *Physoderma alfalfae*. Sorus of resting sporangia in lucerne tissue.

gametothalli are haploid and bear the gametangia. The anisoplanogametes swim and conjugate in water. The zygote then develops into a sporothallus which is diploid and bears sporangia. Some of the sporangia are thin-walled and produce diploid zoospores which can germinate to form new sporothalli. Other sporangia, however, are thick-walled resting sporangia in which meiosis occurs and results in the formation of haploid zoospores which, on liberation, germinate into new gametothalli. In *Allomyces* the gametes are guided to one another by a chemotropism involving a hormone called **sirenin** produced by the female gametes. The degree of maleness or femaleness of the thalli can be controlled experimentally by altering the chemical composition of the medium.

Order Monoblepharidales

Few species belonging to this order are known and they are mostly aquatic. In the genus *Monoblepharis* the eucarpic thallus consists of well branched, highly vacuolated hyphae whose cytoplasm appears foamy. The sporangia are elongated and cylindrical, borne at and near the apices of hyphal branches. They produce zoospores which swim for a time, encyst, and then germinate by a germtube to form a new mycelium. Gametangia are formed on the same thalli in response to the onset of higher temperatures. The antheridia, containing the male antherozoids, are borne laterally on the larger oogonia. Antherozoids are liberated and swim to the oogonium, passing through a papilla in its wall and fertilizing the uninucleate oosphere. The diploid zygote emerges from the oogonium and becomes attached to the latter by a collar; there it develops further into a thick-walled, warted oospore which is able to germinate into a new mycelium. Among fungi, such fertilization of a non-motile female gamete by a male planogamete is unique to *Monoblepharis*.

CLASS OOMYCETES

Saprophytism and the aquatic habitat are usual in the lower Oomycetes, especially the 'watermoulds' (Saprolegniales), but the more advanced 'downy mildews' and 'white rusts' (Peronosporales) are obligate parasites of land plants and are predominantly disseminated by wind. The majority of Oomycetes are eucarpic. Asexual reproduction predominates and there may be several asexual cycles in a favourable season. Zoospores are produced in almost all Saprolegniales and in all but the most advanced types of Peronosporales where the sporangium may function as a conidiosporangium with a purely conidial function at times. Sexual reproduction is less varied than in the Chytridiomycetes, almost always being heterogamous, with gametangial contact resulting in the formation of one or more oospores as the zygote. The oospores are resistant and appear to be a means of surviving poor conditions.

Order Saprolegniales

The watermoulds occur typically in fresh water or in wet soil, as in ditches. Most are saprobic, but certain species of *Saprolegnia* parasitize fishes and their eggs, while species of *Aphanomyces* are important as root pathogens of some crop plants. The somatic hyphae in members of the Saprolegniales range in width from narrow to very wide indeed and are attached by rhizoids. Irregular terminal or intercalary segments of hyphae may be cut off as **gemmae** with dense cytoplasm when conditions are unfavourable for production of zoospores; these function like chlamydospores.

Sexual reproduction occurs by gametangial contact and fertilization of one or more oospheres within the oogonium; male nuclei pass into the oogonium through fertilization tubes put out by the antheridial branches (Fig. 32; *Achlya*). The zygotes are smooth, thick-walled oospores which may either develop into a zoosporangium or germinate by a germtube. In *Achlya* at least four hormones control the development of the sex organs at different stages; female hormones initiate the development of the antheridia, whose branches in turn produce male hormones which stimulate the production of oogonia, and so on.

In the asexual phase of Saprolegniales there is a tendency for the swarming periods of zoospores to be suppressed until finally, non-motile sporangiospores are evolved. There is also a tendency for the sporangium (a spore mother-cell) to become deciduous and to assume the dispersal function of a spore.

The sporangia of the Saprolegniales do not differ greatly in appearance from cells of the wide hyphae from which they are formed, i.e. they are more

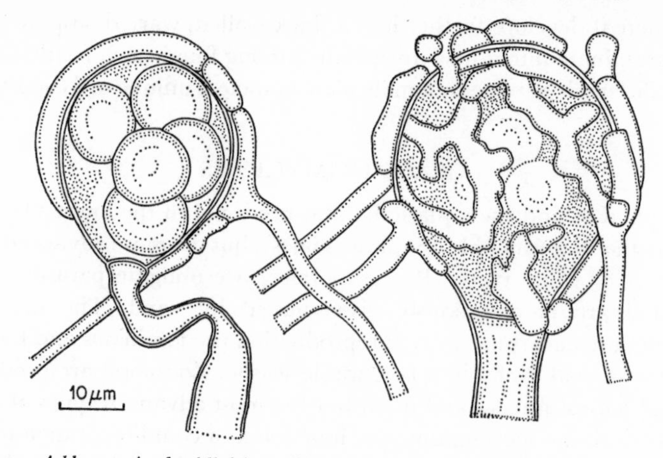

Fig. 32. *Achlya* sp. Antheridial branches surrounding oogonia containing oospores (left) and oospheres (right).

or less cylindrical and thin-walled. Although intercalary sporangia may sometimes occur, terminal ones are more typical; they open at maturity by a simple exit pore at the apex. In *Saprolegnia* and *Achlya* the sporangia remain attached to the thallus after the zoospores have been liberated, and soon proliferate new sporangia, but in different ways. In *Saprolegnia* the new sporangia proliferate internally from the base of the old sporangium and within its empty sac, a process which may be repeated several times. In *Achlya* proliferation takes place by cymose branching, each new sporangium developing at the apex of a branch which originates laterally from below the basal septum of the previous sporangium in the succession (Fig. 33).

Zoospores of two morphological types may be produced. Both types are biflagellate with whiplash and tinsel flagella, but the **primary zoospores** are pyriform with apical flagella while the **secondary zoospores** are reniform with the pair of flagella lateral on the concave side. A species may thus be said to have **monomorphic** or **dimorphic** zoospores according to whether only one, or both types are present in the life-cycle. Zoospores of both types may show different patterns of swarming behaviour. **Monoplanetic** zoospores have only one period of swarming before rounding off and germinating by a germtube. With **diplanetic** zoospores there are two motile periods separated by a period of encystment; and with **multiplanetic** zoospores several motile periods are separated by encystment. By extension, it is customary to speak of monoplanetic, diplanetic and multiplanetic species, though this usage is not strictly correct. Encystment of the zoospores must be of biological advantage in unfavourable conditions, e.g. temporary drought. What is puzzling, however, is the change in form from primary to secondary type of zoospore, at least in many species.

In the genus *Pythiopsis* only primary zoospores are formed, but in some

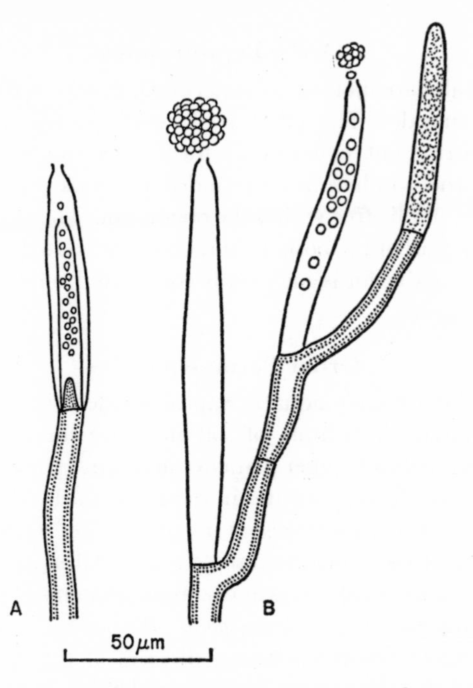

Fig. 33. Sporangia showing method of proliferation. A, *Saprolegnia* sp.; B, *Achlya* sp.

species they may be monoplanetic and in others diplanetic. In *Saprolegnia* the zoospores are dimorphic and diplanetic, with the swarming periods of both primary and secondary zoospores lasting a long time. In *Achlya* the first swarming period is reduced in duration, the primary zoospores encysting just outside the exit pore of the sporangium. In *Dictyuchus* the first swarming period is entirely suppressed and the primary zoospores encyst while still within the sporangium. As a result of close packing in the sporangium they assume a polyhedral shape and thus the sporangium falsely appears to be multicellular. The primary zoospores may then either germinate directly by a germtube which emerges through the sporangial wall, or may release secondary zoospores through exit pores; such secondary zoospores are multiplanetic. The whole sporangium of *Dictyuchus* is deciduous and is dispersed from the thallus. Finally, in *Geolegnia*, all swarming is suppressed entirely and no zoospores are formed; instead the sporangiospores are thick-walled, enlarged, aplanate and germinate by a germtube. This suggests that aplanate sporangiospores, as typical of the Zygomycotina, might have evolved from a type like *Geolegnia* by elimination of the swarming periods of zoospores. Although the sporangiospores of *Geolegnia* are aplanate, this genus has other features which align it with the Oomycetes rather than with the Zygomycotina, notably its method of sexual reproduction.

Order Leptomitales

The Leptomitales are more or less intermediate between the Saprolegniales and the Peronosporales. This order is of interest chiefly because: the sporangia are pyriform, not cylindrical; there is a single oosphere in each oogonium; and the oosphere is surrounded by thick protoplasm, the **periplasm,** and does not lie free within the oogonium. The periplasm eventually forms part of the wall of the oospore and becomes warted, spiny or reticulate. The hyphae are notable for being deeply constricted at intervals, the constrictions simulating septa.

Order Peronosporales

In contrast to the largely aquatic Saprolegniales, the Peronosporales are found as saprophytic inhabitors of soil moisture films from which many species may invade roots of higher plants as facultative parasites; alternatively, they may occur as obligate parasites in the aerial parts of higher plants. The Pythiaceae (e.g. *Pythium*; *Phytophthora*) invade roots of seedlings from soil and cause root-rots and damping-off diseases. The Albuginaceae (e.g. *Albugo*) cause 'white rust' diseases of leaves and sometimes stems of herbaceous plants. The Peronosporaceae (e.g. *Peronospora*, *Plasmopara*, *Bremia*, *Sclerospora*, *Basidiophora*) produce 'downy mildew' diseases of the leaves and stems of higher plants. All these organisms cause considerable economic loss in crops.

The sexual states in all Peronosporales are very similar. Oogonia and antheridia (Fig. 34) are borne on the same or on different hyphae. The oogonium usually contains a single oosphere surrounded by periplasm. Gametangial contact occurs, with the formation of a fertilization tube, and

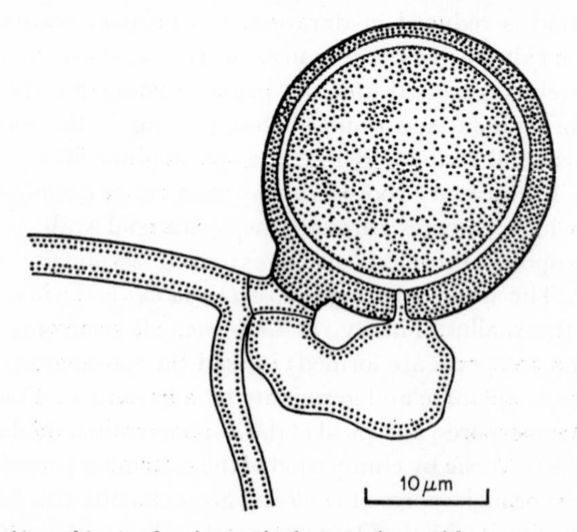

Fig. 34. *Pythium ultimum*, showing a single oogonium with oospore, antheridium and fertilization tube.

the resulting zygote is a thick-walled oospore which is usually ornamented. It germinates either by producing zoospores (thus behaving as a resting sporangium) or by a germtube which gives rise to a sporangium.

In the Peronosporales one can trace a series of examples in which there is a tendency to suppress zoospore-formation and produce conidiosporangia which function facultatively as conidia. In all families of the Peronosporales examples are to be found in which the sporangia are deciduous, dispersed as whole units, and then differentiate their contents into zoospores which are liberated and germinate by germtubes. In other conditions of moisture and temperature, however, the contents of the sporangia do not cleave into zoospores; the sporangium is still dispersed as a single whole cell, but it germinates directly by a germtube and is essentially conidial in function and form. This function is most obvious in the leaf-inhabiting parasitic Peronosporales, and the sporangial function is more typical of the soil and root inhabitors. This is a clear example of the importance of habitat in determining development and functions in fungi. The biological advantages of being able to vary the spore type to fit the circumstances are fairly obvious.

Family Pythiaceae

These are the damping-off and root-rot organisms. Members of the genus *Pythium* are interesting not only for their pathogenic potential but also for the variety of sporangia formed. These are produced on fine, profuse mycelia, usually with no distinctive sporangiophores. The simplest sporangia, as in *Pythium gracile*, are cylindrical and can hardly be distinguished from ordinary hyphal apices; they liberate zoospores held together for a time in a cluster at the open end of the sporangium. In most species of *Pythium* (e.g. *P. debaryanum*; *P. coloratum*) the sporangium is spherical, with a firm peridium. At maturity it gives rise to an exit tube of variable length, at whose apex an extremely thin-walled vesicle is formed (Fig. 35). Protoplasm enters the vesicle and becomes differentiated there into zoospores which are liberated as secondary zoospores and germinate by a germtube. In *P. intermedium* the sporangia are borne in basipetal chains, a feature which is reminiscent of the Albuginaceae. In still other species of *Pythium* (e.g. *P. ultimum*) the sporangia may assume an entirely conidial function.

The genus *Phytophthora* contains species which are very similar to species of *Pythium* but differ in that no vesicle is formed from the sporangia, or if one is formed then the zoospores enter it in a mature condition. Such a distinction is certainly difficult to recognize in practice, especially in view of the capacity of sporangia for variation. In the many soil-inhabiting species of *Phytophthora* no distinctive sporangiophores are formed. An interesting feature is the chemotropic attraction of the zoospores to substances liberated into the soil of the rhizosphere from the rootlets of higher plants; unwittingly, the plant may seal its doom by attracting zoospores of species which are able to

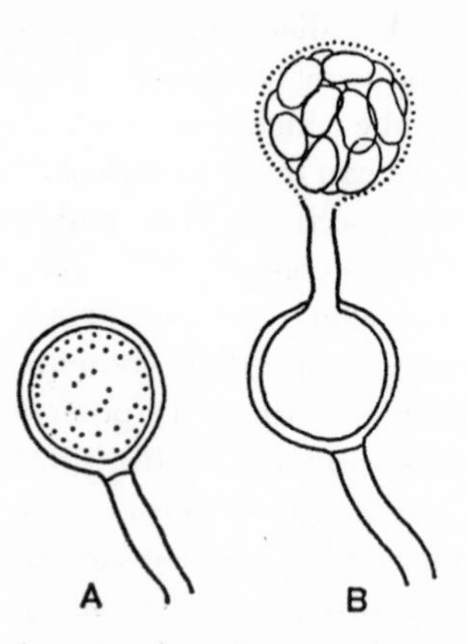

Fig. 35. *Pythium coloratum.* A, sporangium; B, empty sporangium with exit tube and vesicle containing reniform secondary zoospores. (A and B from a photograph supplied by courtesy of Dr O. Vaartaja.)

attack and invade its roots. *Phytophthora infestans*, a pathogen of the aerial parts of potato and the cause of potato late blight disease, differs from the soil-inhabiting species in many respects, but especially in forming sympodially branched indeterminate sporangiophores (Fig. 36) in profusion on the lower surfaces of leaves. The sporangia formed on these are conidiosporangia, with temperature controlling their mode of germination to zoospores or by germ-tubes. This species caused the disastrous Irish potato famine of 1845, whose effect on British politics and mercantile history is admirably traced by Large (1940).

Family Peronosporaceae

The sporangia in members of the Peronosporaceae, or downy mildews, are oval, spherical or limoniform and are borne on small denticles which terminate the branchlets of determinate sporangiophores. The sporangiophores are hygroscopic and, with variation in the moisture content of the surrounding air, undergo twisting movements which are violent enough to dislodge the deciduous, wind-dispersed sporangia. In most instances they are conidiosporangia, but in members of the genus *Peronospora* they always function as conidia.

The various genera of the Peronosporaceae are readily distinguished by the type of branching of their sporangiophores (Fig. 36). In *Peronospora*, whose species are pathogens of various plants, the branching is dichotomous and the ultimate branchlets (denticles) are long, curved and set at an acute angle to each other. In *Pseudoperonospora*, often a pathogen of cucurbits, the branching is similar but the sporangia are coloured. In *Plasmopara* the branching is irregular, at a wide angle, and the denticles are short, straight

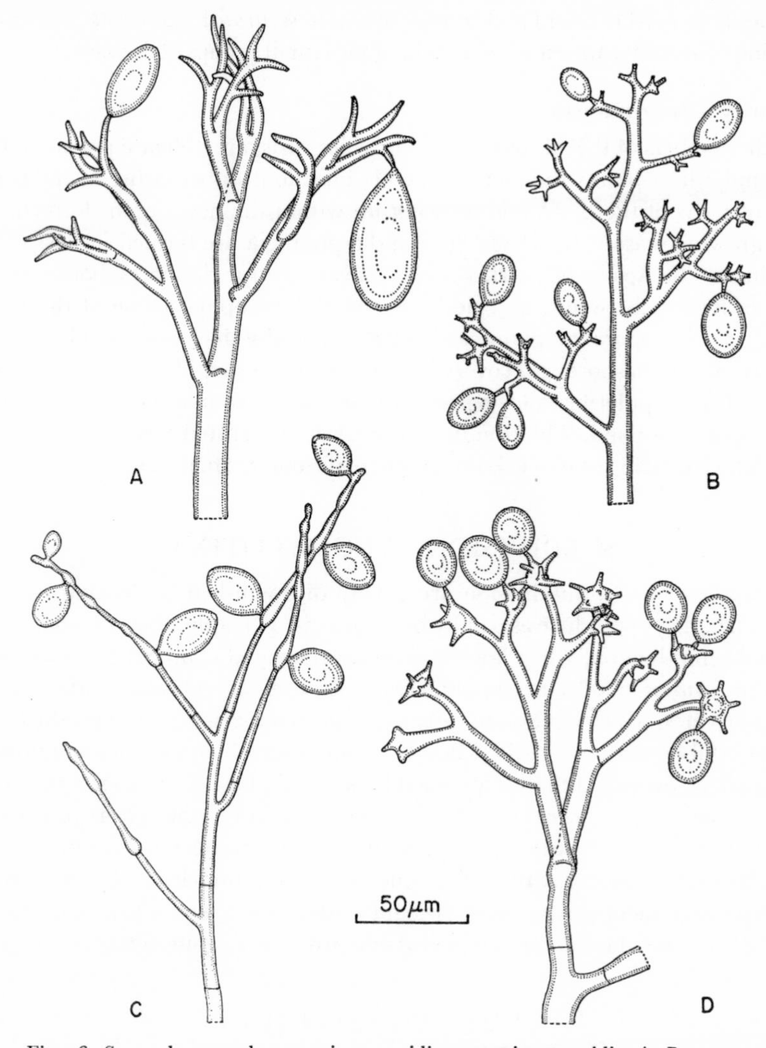

Fig. 36. Sporophores and sporangia or conidiosporangia or conidia. A, *Peronospora destructor*, conidia; B, *Plasmopara viticola*; C, *Phytophthora infestans*, with sympodial indeterminate sporangiophores; D, *Bremia lactucae*. The sporangiophores in A, B, D, are determinate.

and usually in pairs or in threes. *Plasmopara viticola* is a serious pathogen of grape vines. In *Bremia lactucae*, which attacks lettuce, the apices of the branches are expanded into discs bearing several denticles around their margins. *Basidiophora* has a simple, clavate sporangiophore with several small denticles arising from its inflated apex; two species of this genus occur on Compositae. In *Sclerospora*, whose species are pathogens of cereals, the sporangiophore is arborescent and consists of a stout stem with numerous branches arising towards its apex and rebranching into small groups of denticles.

Family Albuginaceae

All members of the Albuginaceae, or white rusts, are obligate parasites. In perennial host plants they form systemic infections, hibernating in the persistent parts of the plant and growing up within the new organs formed in the growing season. In *Albugo* the conidiosporangia are formed in basipetal chains at the apices of short, simple, clavate sporangiophores borne in a **sorus** formed between the epidermis and the mesophyll tissue of the host (Fig. 37). The sporangiophores are characteristically thick-walled. The sorus, a mass of conidiosporangia covered at first by the host epidermis, later bursts through the epidermis and exposes the deciduous conidiosporangia to dissemination by wind. The conidiosporangia are separated from one another in their chains, by a small gelatinous pad of disjunctive material.

SUBDIVISION ZYGOMYCOTINA

Members of this Subdivision are chiefly distinguished by lacking motile cells; i.e., they reproduce asexually by sporangia with aplanospores, and their sexual reproduction, when known, is by gametangial conjugation, resulting in the formation of a zygospore. These features are typical of the Class Zygomycetes, which contains mostly saprobic members. Sexual reproduction is doubtful or unknown in the Class Trichomycetes which includes fungi with a simple or branched thallus attached by a discoid base to the digestive tract or external cuticle of arthropods. The relationship is regarded as probably one of commensalism rather than parasitism. The sporangia in Trichomycetes are linear, containing a row of uninucleate or multinucleate aplanospores. Manier and Lichtwardt (1968) recognize four orders and seven families of Trichomycetes. This account will deal only with the Zygomycetes.

CLASS ZYGOMYCETES

The two orders of Zygomycetes are the Mucorales, whose members are mostly saprobes and reproduce asexually by means of sporangia containing one or more aplanospores, and the Entomophthorales, whose members are mostly entomogenous (on insects) and reproduce asexually by means of violently discharged conidia.

Order Mucorales

While most Mucorales, or pin-moulds, are saprobes a few are parasitic on other fungi, e.g. species of the genera *Piptocephalis* and *Syncephalis*. Assuming that *Piptocephalis* is a parasite only of other Mucorales, Fenner (1932) tested

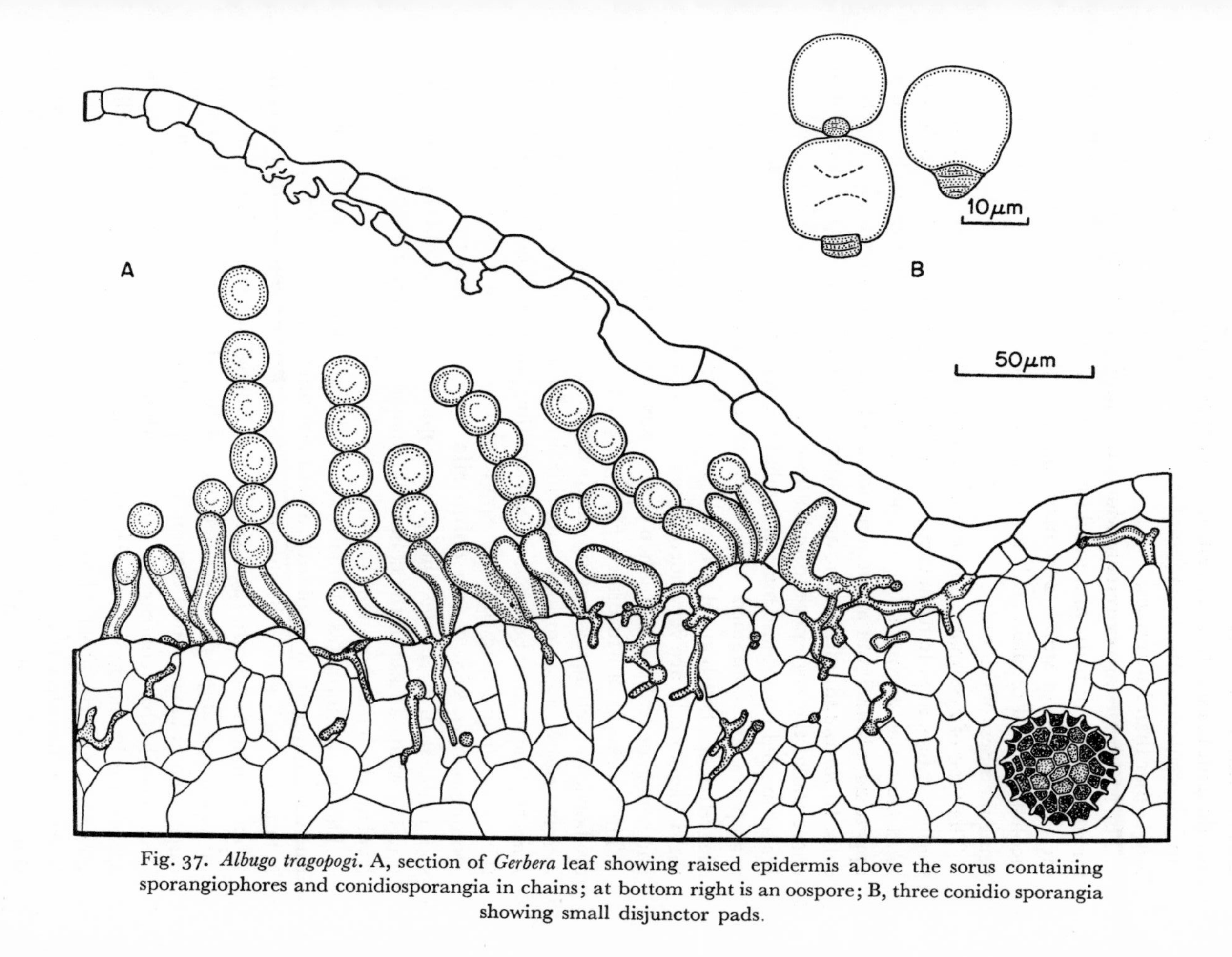

Fig. 37. *Albugo tragopogi.* A, section of *Gerbera* leaf showing raised epidermis above the sorus containing sporangiophores and conidiosporangia in chains; at bottom right is an oospore; B, three conidio sporangia showing small disjunctor pads.

its parasitism on *Mycotypha microspora* in order to provide additional evidence that this unusual conidial fungus is mucoraceous, since its systematic position was difficult to establish in the absence of zygospores. The tests were positive, but it has subsequently been found that species of *Piptocephalis* may parasitize fungi other than Mucorales.

The Mucorales are extremely widely distributed and many are responsible for storage-rots of fruits and moulding of bread. Some are concerned in the saccharification of starch and may continue the fermentation to produce alcohol. Several organic acids are produced commercially by fermentations using members of this order.

The type of zygospore, and the manner in which it may be protected by sterile hyphae, are often characteristic of a genus of Mucorales (Fig. 38). In *Phycomyces* the zygospore is smooth, and stiff, dark brown dichotomous appendages branch out at the junction(s) of the zygospore with its suspensor cells. In *Absidia* simple hyaline appendages arise from the suspensor cells and curve round the zygospore from both sides. In *Mortierella* a single zygospore is surrounded by a dense mass of hyphal branches, while in *Endogone* several zygospores are contained within a common covering of sterile hyphae suggestive of a fruitbody of one of the higher hypogean fungi. In *Zygorhynchus* the suspensor cells are very unequal in size. Zygospores are sometimes formed parthenogenetically and are then termed **azygospores.**

The sporangia of Mucorales (Figs. 39, 40, 41) are globose to pyriform and are borne on well developed unbranched or branched sporangiophores which in some cases arise in groups from rhizoids. In some genera (e.g. *Mucor* and *Rhizopus*) the septum between the apex of the sporangiophore and the base of the sporangium bulges into the sporangium to form a prolongation of the sporangiophore known as a **columella.** In certain genera with pyriform sporangia, such as *Absidia*, the sporangiophore widens gradually and forms a wedge shaped base to the columella, known as the **apophysis,** below the level at which the sporangial peridium is attached (Fig. 39).

The sporangiophores of many Mucorales are positively phototropic but this would appear to be of little advantage unless they were fairly long and able to rise well above the substratum, as in *Phycomyces* where they may be 10 cm or more in length. Ingold and Zoberi (1963) have investigated the ways in which sporangiospores of members of the Mucorales are presented for dispersal and have related them to the types of sporangia present. This work shows clearly that, contrary to what is perhaps a generally held opinion, the spores are by no means all dry and wind-distributed. Large columellate sporangia may be dispersed by wind without dehiscing (*Actinomucor*), or may burst and liberate either dry spores (*Rhizopus* and some species of *Mucor*) or slimy spores held in a mucilaginous droplet (other species of *Mucor*; *Phycomyces*). In Mucorales with reduced sporangia known as sporangiola and merosporangia, the spores are dry in some examples and slimy in others; the same

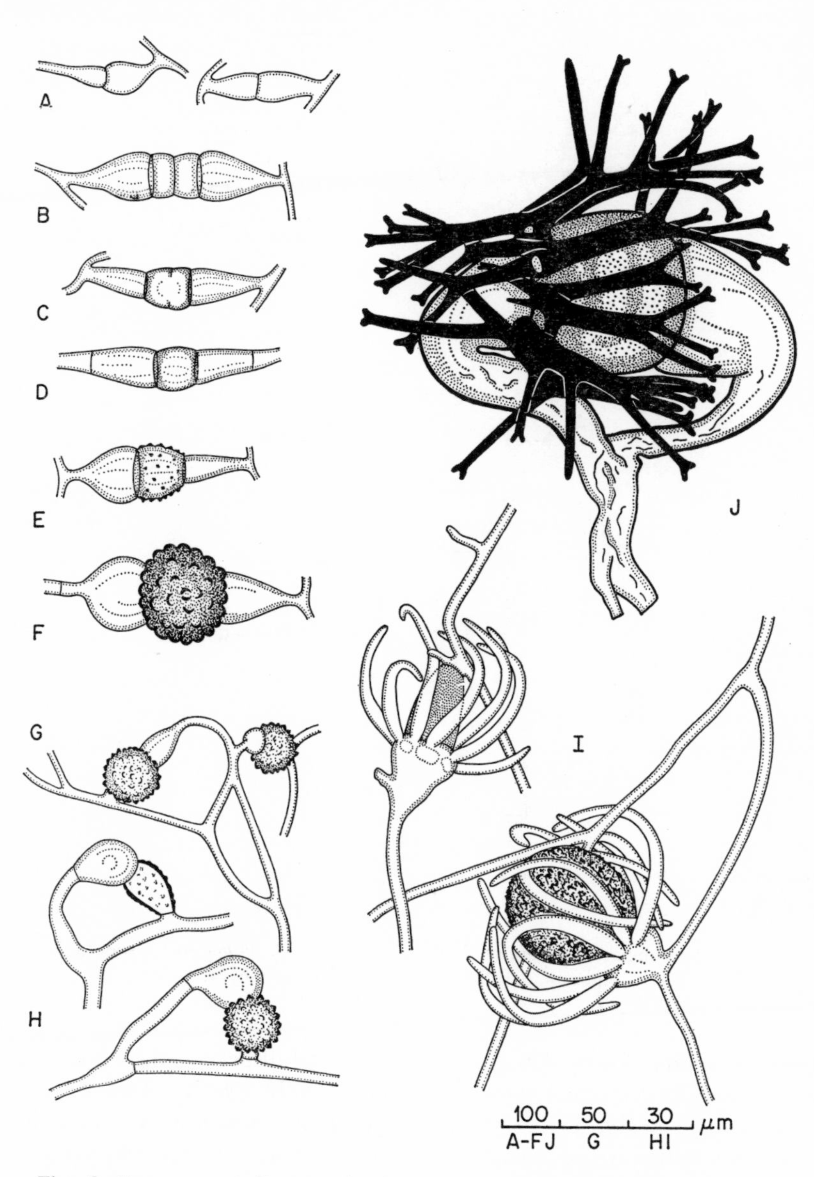

Fig. 38. Zygospores. A–F, stages in plasmogamy by gametangial conjugation to form zygospores in *Rhizopus* sp.; G, H, zygospores and their very unequal suspensor cells in *Zygorhynchus* sp.; I, two zygospores of different age in *Absidia* sp., each surrounded by curved appendages branching from the suspensor cells; J, Zygospore of *Phycomyces* sp., surrounded by stiff dark appendages branching dichotomously from one or both suspensor cells.

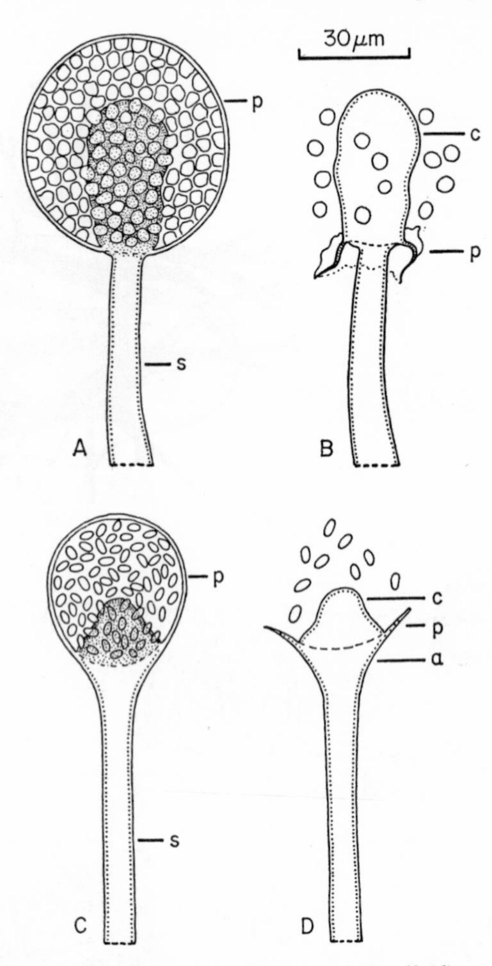

Fig. 39. Sporangia of Mucorales. A, B, *Mucor* sp.; C, D, *Absidia* sp.; *s*, sporangiophore; *p*, peridium; *c*, columella; *a*, apophysis.

holds for conidial forms. The dry spores are wind-dispersed; the slimy spores are presumably dispersed by contact, by insects or by splashes of water.

In *Pilobolus* species (Fig. 40) there is an elegant mechanism of sporangium-discharge (Buller, 1934) which clearly profits from the fact that the sporangiophores are positively phototropic. *Pilobolus* species occur on the dung of herbivores. Their erect sporangiophores are swollen distally into a rather large subsporangial vesicle with an apical columella surrounded by the small black sporangium whose peridium is somewhat thickened and water-repellent on the distal side. The swollen vesicle acts as a convex lens. Its apex is shaded by the dark sporangium, so that when it is directed accurately

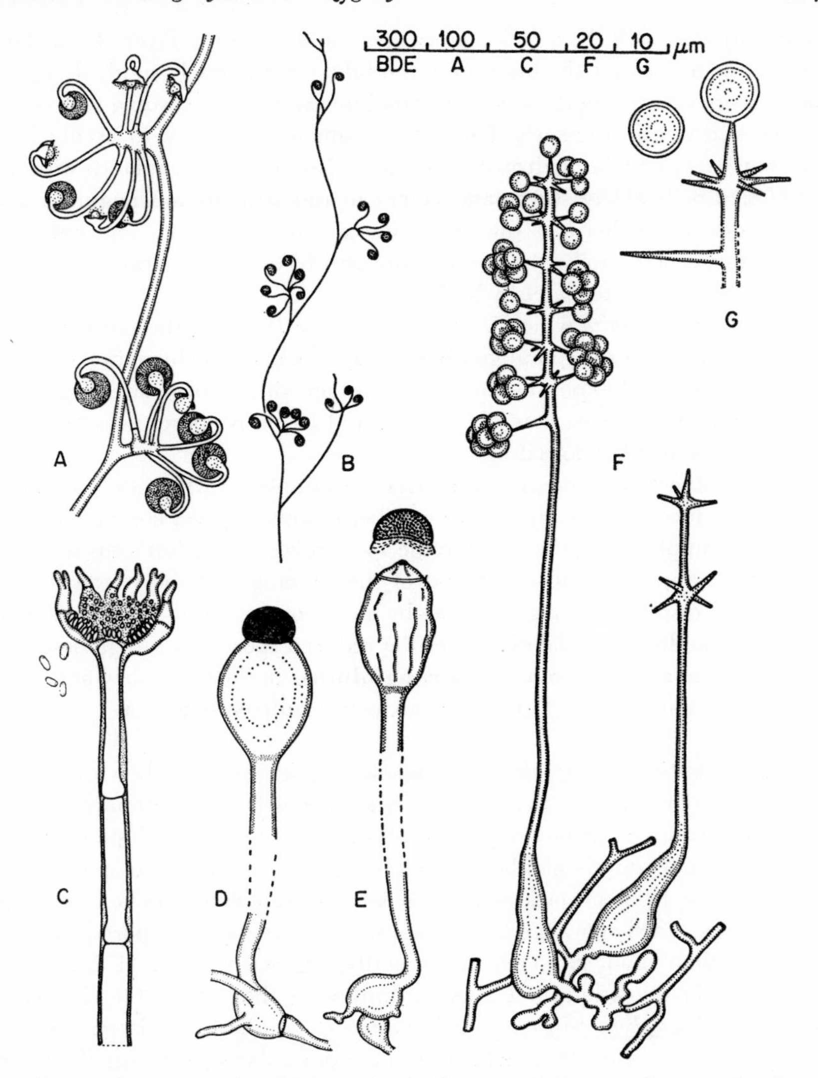

Fig. 40. Some examples of Mucorales. A, B, sporangiophores and sporangia of *Circinella* sp.; C, sporangiophore, sporocladia, pseudophialides and merosporangia of *Kickxella alabastrina*; D, E, *Pilobolus* sp., sporangiophore arising from a basal trophocyst and expanding at apex into the subsporangial vesicle bearing a dark sporangium; F, G, *Mortierella* sp., conidiophore and conidia.

towards a light source the light rays are focused near the base of the vesicle. When light rays come from an angle slightly to one side, they are focused some distance up the opposite side of the vesicle, which then grows unilaterally and again turns the sporangiophore accurately into the beam of light. At maturity the sporangial wall ruptures circularly around the level of its junction with the vesicle, gapes open, and exposes a layer of mucilage on its under side. At the same time the fully turgid vesicle suddenly bursts near its apex, carrying the whole sporangium away in the ensuing rush of sap under hydrostatic pressure. The sporangium is shot away to as much as 180 cm, carrying a large drop of sap with it. No matter how it lands with the drop of sap, it always turns and comes to rest with its mucilaginous side facing downward because the other side is water-repellent. The mucilage dries quickly and sticks the sporangium firmly to the substratum, usually herbage which is eaten later by herbivores.

In the genus *Pilaira* which is superficially like *Pilobolus*, the subsporangial vesicle is small but the sporangiophore is capable of great elongation towards a source of light; the sporangium makes contact with herbage after rupturing in the same manner as that of *Pilobolus*, and its slimy spores adhere to the herbage in a mucilaginous drop.

In the Mucorales there are two series of examples in which the sporangia become reduced in size and in the number of aplanospores contained, until finally, examples are known where the sporangia are monosporous with the spore wall apparently fused to that of the sporangium; the whole is then deciduous, dispersed, and germinates by means of a germtube, thus functioning as a conidium. In the one series the reduced sporangia are globose and are known as **sporangiola** (s. **sporangiolum**); in the other they are cylindrical with a linear row of spores and are called **merosporangia.**

(a) Series with sporangiola: Sporangiola (Fig. 41) are well shown in members of the genus *Thamnidium* where they are formed on the ultimate branchlets of determinate, dichotomous lateral branches of a sporangiophore. They may arise from a sporangiophore which bears a large columellate sporangium terminally, or from one lacking this; the large sporangia tend to be suppressed in darkness. Each sporangiolum contains only a few aplanospores, perhaps not more than about six to ten. Incidentally, *Thamnidium elegans* is concerned in a process patented in the U.S.A. for improving the taste and tenderness of beef. In *Choanephora trispora* (=*Blakeslea trispora*) there are large primary sporangia each with a columella and numerous aplanospores, smaller intermediate sporangia without a columella and with fewer aplanospores, and sporangiola each containing only three aplanospores. The sporangiola are borne on minute denticles arising from the swollen apices of branches of the sporangiophore. The process of reduction reaches a further stage in some other species of *Choanephora*, where the sporangiola are monosporous. In *Cunninghamella* the monosporous sporangiola are borne in a manner exactly like the trisporous sporangiola of *Choanephora trispora*, but no primary or intermediate sporangia are present; the sporangiola of *Cunninghamella* are in effect conidia (blastospores) of a type very generally found in higher fungi, and they are disseminated and germinate like conidia. Among the higher fungi, the

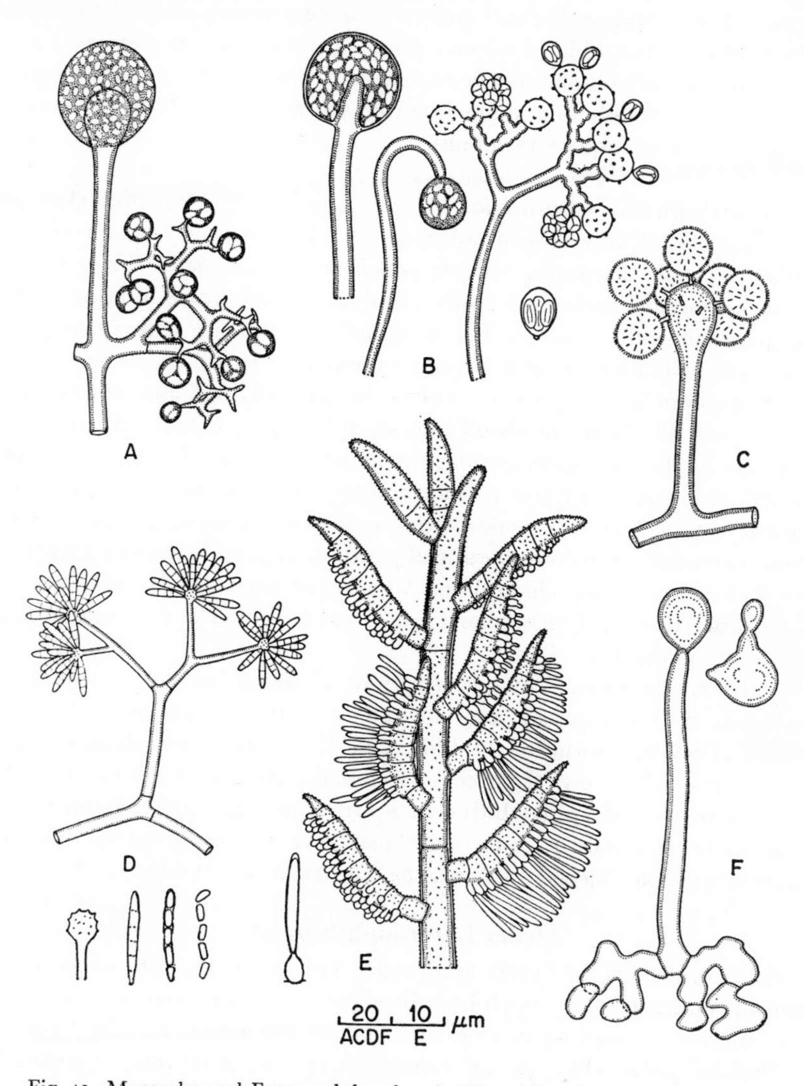

Fig. 41. Mucorales and Entomophthorales. A, *Thamnidium elegans*, primary sporangium with columella; sporangiola borne on dichotomous branches of the sporangiophore; B, *Choanephora* (*Blakeslea*) *trispora*, primary sporangium with columella, intermediate sporangium and sporangiola; C, *Cunninghamella* sp., sporangiophore and monosporous sporangiola or conidia; D, *Piptocephalis* sp., dichotomous sporangiophore with terminal vesicles and radiating merosporangia (above); detail of vesicle and merosporangia (below); E, *Coemansia* sp., apex of sporangiophore with lateral sporocladia bearing pseudophialides and monosporous merosporangia; a single pseudophialide and merosporangium (left); F, *Conidiobolus* sp., 'hyphal bodies', conidiophore and conidium; conidium germinating by repetition (right). (B, based on Thaxter, *Bot. Gaz.*, 1914, **58**, 353.)

formgenus *Oedocephalum* is characterized by having conidiophores and conidia which closely resemble the corresponding structures in *Cunninghamella*. Since the walls of conidia may be two-layered it is difficult to judge whether a structure is a conidium or a monosporous sporangiolum whose wall is fused to that of the aplanospore it contains.

(b) Series with merosporangia: The merosporangiate Mucorales (Fig. 41) have been beautifully monographed by R. K. Benjamin (1959, 1966). In genera such as *Syncephalis*, *Syncephalastrum* and *Piptocephalis*, the apices of sporangiophore branches are swollen and bear dense heads of linear merosporangia on small denticles. In *Piptocephalis* the delicate sporangiophore is branched dichotomously with exquisite regularity. In young merosporangia the protoplasm is homogeneous; later it becomes divided into separate segments, each of which rounds off into an aplanospore contained within the peridium of the merosporangium in a linear row. The number of aplanospores is always relatively low, perhaps as many as fifteen, and in some instances is reduced to only one. A monosporous merosporangium is, in effect, a conidium. Eventually, the merosporangial peridium degenerates and the aplanospores then look like a chain of exogenous spores, or conidia, and are not easily differentiated from similar formations of conidia to be found among the Deuteromycotina.

In the Kickxellaceae (Linder, 1943; R. K. Benjamin, 1959) the merosporangia are monosporous and elongated, or shorter and ovoid in some genera. They are borne on bud-like denticles (**pseudophialides**) which arise in brush-like rows from cells of special fertile branches of the sporangiophore termed **sporocladia** (s. **sporocladium**). These fungi which occur quite commonly on dung of various types, are among the most beautiful of microscopic objects (e.g. *Coemansia*, Fig. 41; *Kickxella*, Fig. 40).

Order Entomophthorales

Most members of the Entomophthorales (Fig. 42) are parasitic on insects or saprobic on insect skins which have been cast, or animal excreta. *Entomophthora* species are well known on flies and on the larvae of other insects; *Conidiobolus* species (Fig. 42) are sometimes isolated from insect remains in soil.

The mycelium in Entomophthorales is reduced and has a marked tendency to septate into small irregular portions known as 'hyphal bodies'. These can conjugate to produce zygospores. They also divide or bud and eventually give rise to conidiophores, each with a single apical **conidium.** The conidiophores are positively phototropic and discharge the conidia violently towards the light source, a mechanism which presumably has some biological advantage in ensuring that the conidia will be dispersed away from the low-lying substratum and will enter the layer of tubulent air above it.

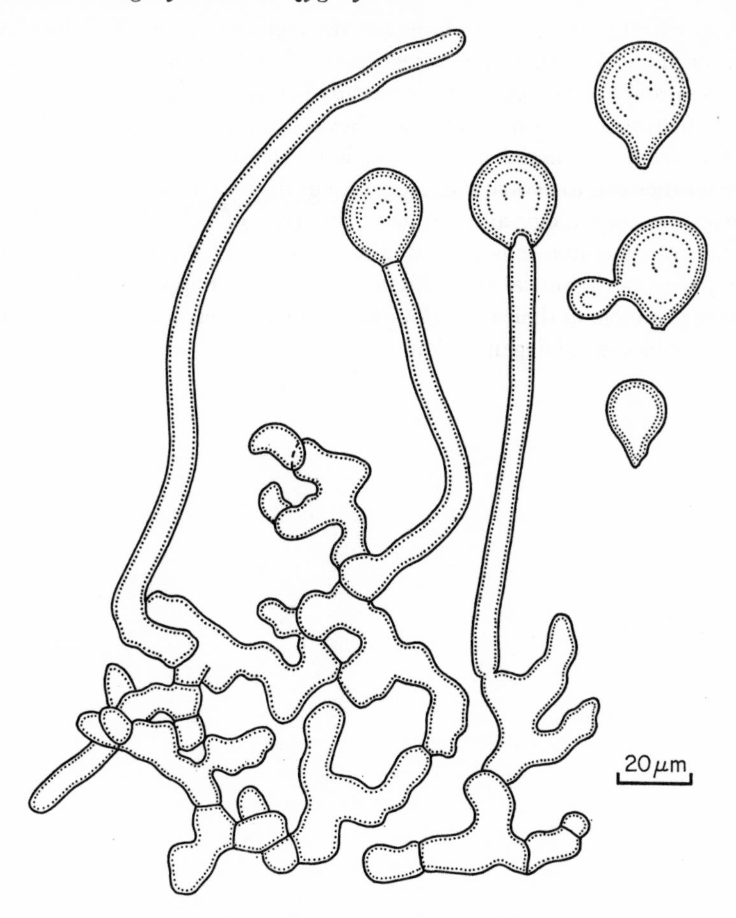

Fig. 42. *Conidiobolus* sp., conidiophores arising from segmented mycelium ('hyphal bodies'), conidia, and repetitive conidium.

The liberation mechanism in *Conidiobolus* depends on rounding-off of a turgid cell; the apex of the conidiophore at first invaginates a thin-walled basal portion of the conidium which suddenly becomes everted when the conidium is fully turgid, with sufficient force to throw the conidium into the air. That these conidia represent monsporous sporangiola can be shown by soaking them in water, when they may exhibit an outer peridial wall and a single 'spore' within it. The conidia germinate in most instances after undergoing **repetition.**

In *Basidiobolus*, species of which occur on excreta of frogs and lizards and may be pathogenic to humans, the conidium and the upper part of the conidiophore are violently discharged together. The swollen conidiophore ruptures towards its base and the sap it contains squirts back so that the recoil carries the upper part of the conidiophore and the conidium away. This mechanism has been likened to that of a water-rocket (Ingold, 1961). Some features of growth in *Basidiobolus ranarum* have already been noted (p. 49).

One feature common to all the above examples of gradual change in function from sporangial to conidial, is the tendency for a non-deciduous spore mother cell to evolve into a deciduous dispersed spore, with correlated changes in morphology and habitat. This type of behaviour, in which the function of daughter cells is gradually assumed by the mother cell, is also clearly seen in the sexual reproduction of fungi and appears to have been a guiding principle in their evolution. Despite its success it is often regarded as indicating biological degeneration.

10

FUNGI WITH A STERILE MYCELIUM OR CONIDIA: EUMYCOTA SUBDIVISION DEUTEROMYCOTINA

STERILE MYCELIA

Not all Deuteromycotina are found in a sporing condition; some species of great economic importance occur regularly as sterile mycelia on natural substrata and remain sterile when isolated in culture, or may be induced to sporulate only by hit-and-miss variation of cultural conditions. The factors inducing sporulation differ from one species to another and are not generally understood, so it is not surprising that attempts to induce sporulation are empirical, not always successful, and are very often not pursued. Yet some sterile mycelial states, which may or may not include sclerotial states, are important enough to require naming and classification and are sometimes distinctive enough to be described and recognized again with reasonable confidence.

Racodium cellare, a sterile mycelium in wine cellars, has profusely warted, brown or black hyphae with septa typical of a member of the Ascomycotina; however, it is probably recognized as much by its habitat as by its morphology. *Ozonium auricomum,* a distinctive golden-brown mycelium forming tufts and strands in litter and on logs, is known to be the mycelial state of a species of *Coprinus,* a mushroom. The sheet-mycelium of various polypores, occurring between bark and wood of trees, has been described under the name *Xylostroma.* Various species of the form-genus *Sclerotium,* a sclerotium-bearing imperfect state, are important pathogens of many plants including rice, ornamentals and onion; some of these have been shown to be imperfect states of certain species of Ascomycotina and Basidiomycotina. The form-genus *Rhizoctonia,* characterized by its rather wide hyphae, with wide-angled branching and with the lateral branches narrowed and septate near their junctions with the main axis, comprises about sixty described species, many of which are notable pathogens with a very wide host range, but also able to exist saprobically. Perfect states of some of these species have been obtained and have been found to be distributed over a wide range of Basidiomycotina and Ascomycotina; some form mycorrhizas with the underground organs of orchids (Warcup and Talbot, 1966, 1967).

Although it is legitimate and sometimes desirable to give names to sterile mycelia, it is preferable at least to try to induce them to form spores so that they can be more accurately classified. One reason for this is the difficulty of describing mycelia in specific terms; and arising out of this is the tendency to broaden generic concepts until they become almost meaningless. For example, several species with hyaline hyphae have been included in *Rhizoctonia* simply because the hyphae have the type of branching and septation described for the type species of *Rhizoctonia*; but it is quite clear that this is a very general type of branching and has little generic significance. Moreover, the hyphae in the type species of *Rhizoctonia* are purple-brown and very different from the yellow to brown pigment in hyphae of most species assigned to *Rhizoctonia*.

CONIDIA AND CONIDIOPHORES

It has already been noted that conidia and other spores can be classed according to their colour and septation into various types which have descriptive value (Saccardo's terminology, p. 81). But these groupings have less taxonomic value. For example, conidia of different age in the same specimen may show differences in colour and septation; spring and autumn collections of the same fungus may differ in the presence or absence of dark pigment (S. J. Hughes, pers. comm.). A single thallus may produce more than one type of asexual spore, thus posing nice problems in taxonomy and nomenclature.

Because convergent evolution is so common in fungi it is possible for similar end products (similar conidia in this case) to be formed in several different ways, and in unrelated organisms. Any taxon based on similarity of the mature conidia is likely to be more heterogeneous than if development had been taken into account and the several different ways of forming the conidia had been recognized. Following the earlier work of Vuillemin, Mason (1933, 1937, 1941) and Hughes (1953) studied the development of conidiophores and conidia in a very wide range of Deuteromycotina, and were able to distinguish several types of conidia according to their methods of formation and their relationships with the conidiophores, disregarding in the first instance, similarities of colour, septation and aggregation of the conidiophores and conidia. Tubaki (1958) and Subramanian (1962), have added a few more general types, but the remarkable thing is that conidia in general are formed in so few ways. Thus ontogenetic typification of conidia offers great possibilities in taxonomy, not merely because there are few major ontogenetic types, but also because these are based on precise and stable features in contrast to the indefinite and variable ones stressed in Saccardo's system.

New precision in description is now possible. Genera based on perfect states of Ascomycotina and Basidiomycotina, but composed of species with totally unlike imperfect states are now at least suspect and require re-evaluation. To date, the best analysis and summary of conidial ontogeny in the Deuteromycotina is that given by Madelin (1966).

Conidiophores

Conidiophores are simple or branched hyphae with the special function of producing conidia; in form, they may or may not be greatly different from somatic hyphae.

In most instances elongation of the conidiophore and initiation of the conidia are confined to the apical region of the conidiophore; however, in a **basauxic** conidiophore growth in length is restricted to a basal cell or cells with which conidium-production is also associated. The apical zone of differentiation of a conidiophore may give rise to a single conidium or, more often, to a succession of conidia. Successive conidia, if dry, tend to adhere in a chain; if slimy, in a droplet of mucilage. But large dry conidia often fall away singly as soon as they are mature. The succession of conidia is commonly **basipetal,** with the oldest conidium at the apex of the chain, but less often **acropetal,** with the youngest conidium at the apex. The formation of chains or slime-balls of conidia is often associated with virtual cessation of elongation of the conidiophore. When they continue to elongate during the process of forming conidia, at least two distinctive types of conidiophore may be concerned: annellophores and sympodulae.

An **annellophore** (Fig. 45, C, D) may be a short sporogenous cell or a longer hypha-like conidiophore but in any case is recognized by the presence of collar-like annellations near its apex. After each, solitary, apical conidium is set free, either by splitting away along the septum at its base or by splitting of the lateral wall of a **separating-cell,** another conidium is formed by proliferation from the conidiophore apex through the scar left by the previously discharged conidium. Successive proliferation thus leaves a series of annellations below the point of attachment of the latest conidium. In at least one species with this type of conidiophore (*Spilocaea pomi*, the cause of apple scab disease), the action of wind-driven rain is necessary to dislodge and disperse the conidia.

A **sympodula** (p. **sympodulae**) (Figs. 44, E, F; 45, A, F) is a conidiophore with subterminal proliferation, which forms an apical conidium, continues its distal growth from a new active area developed to one side of the conidium, and then forms another apical conidium, the process being repeated in succession. The formation of new active areas and conidia successively on alternate sides of the apex results in a zig-zag conidiophore with prominent bends (geniculations) at the points of attachment of the older, now lateral, conidia (Fig. 44, E). If the active areas are formed successively on the same side of the apex, the curious cockscomb-type of conidiophore sometimes called a **cervix** (p. **cervices**) results (Fig. 44, F). Finally, if the successive active areas are irregularly distributed in relation to the apex, the mature conidia and their geniculate points of attachment appear randomly distributed around the conidiophore (Fig. 48, B).

A further distinctive type of conidiophore is the **phialide** (Fig. 46). This is a flask-shaped, cylindrical or subulate cell of limited growth, which abstricts conidia in basipetal succession from its apex either by budding or by the formation of transverse septa. Madelin (1966) has drawn attention to four different sorts of phialides distinguished by whether their conidia are formed by budding or by transverse septation, and whether they secede by transverse splitting of septa or by a combination of transverse and periclinal splitting of the walls.

Conidia

The distinction has already been made between thallospores and conidiospores (p. 87), together comprising conidia.

Thallospores

(*a*) ***Arthrospores:*** these (Fig. 43) arise by close septation of the distal part of a hypha, the septa being formed in basipetal succession. Proceeding from the apex to the base of the hypha, each cell in succession becomes rounded off and is eventually set free as an independent arthrospore. The arthrospores secede either by splitting of the transverse septa between them, or, if endogenous, by circumscissile splitting of the outer, hyphal wall; in the latter case small pieces of the hyphal wall may persist as projections at the ends of the arthrospores.

(*b*) ***Chlamydospores:*** these (Fig. 43) are formed from solitary or neighbouring intercalary cells of a hypha which round off, usually enlarge, and develop a thickened, often pigmented wall and dense contents comprising food reserves; thus they are able to function as resting spores or survival structures. Chlamydospores do not secede from the parent hypha but instead remain as viable units when the hypha decays. The mature spore is usually more or less spherical or oblong and wider than the supporting hypha. Occasionally, chlamydospores may be formed from the cells of other kinds of spores instead of from hyphae, e.g. from the cells of conidiospores

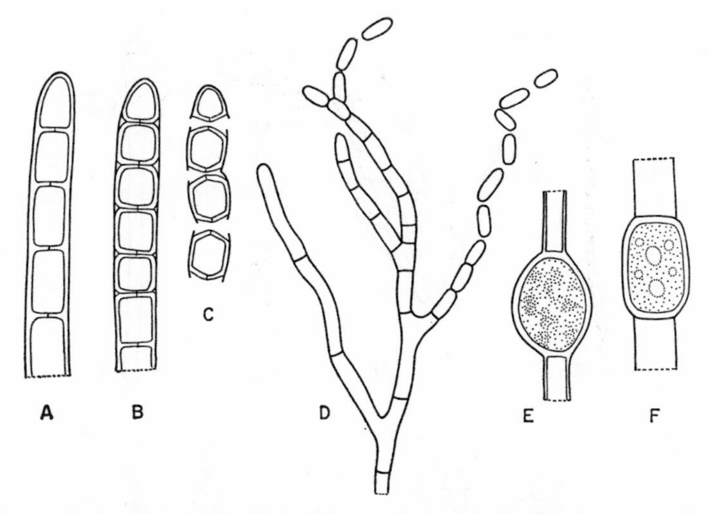

Fig. 43. Arthrospores and chlamydospores (*camera lucida* drawings at various magnifications). A, B, C, *Coremiella ulmariae*, endogenous arthrospores liberated (C) by circumscissile splitting of the hyphal wall; D, *Oidiodendron* sp., arthrospores liberated by transverse splitting of separating septa; E, *Trichoderma viride*, chlamydospore; F, *Mucor* sp., chlamydospore.

in the genus *Fusarium*, or in ascospores of some *Mycosphaerella* species. Depending upon the number of cells transformed, the chlamydospores may be solitary or borne in short chains. In some species of *Fusarium* in soil, the conidiospores and any hyphae and germtubes present are quickly lysed by soil organisms; but the chlamydospores present in the conidiospores are resistant to lysis, remaining dormant until induced to germinate by the presence of exudates from roots of susceptible plants nearby (Cook & Snyder, 1965).

Conidiospores

(a) Blastospores: Blastospores (Fig. 44) or bud-spores, are formed by budding of somatic cells of a hypha or conidiophore, or by budding from the cells of other types of spores. Each bud first appears as a small globular outgrowth from the parent cell and then enlarges while still attached to the parent, finally seceding by abstriction to form an independent spore (or a new individual thallus in some fungi with unicellular thalli, e.g. the yeasts). Acropetal chains of blastospores may be produced by successive budding at the apex, and such chains may be branched since more than one bud may be formed from any spore in the chain. On the other hand, it is also possible for blastospores to arise in basipetal sequence by successive budding from the apex of the conidiophore.

In some species blastospores are formed singly from short denticles arising

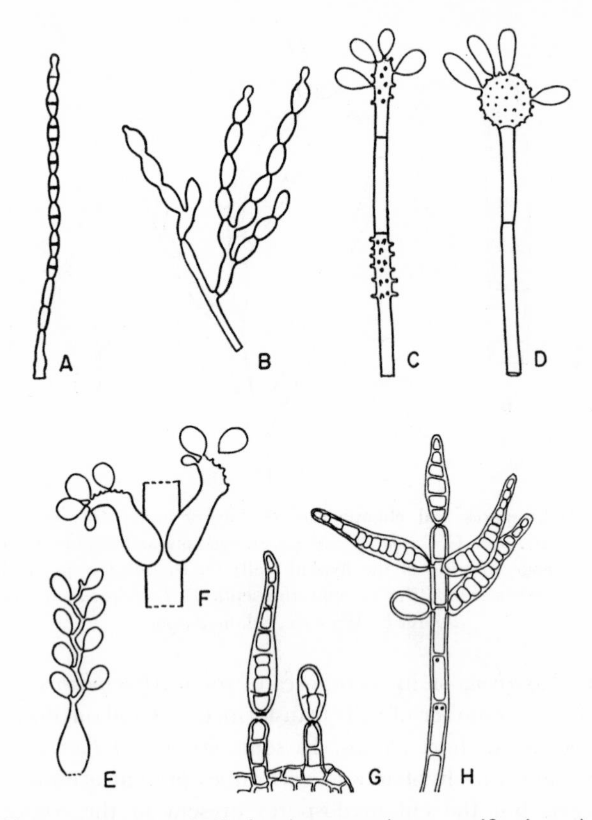

Fig. 44. Blastospores: *camera lucida* drawings at various magnifications. A, *Bispora antennata*, catenate (in chains); B, *Monilinia fructicola*, catenate; C, *Gonatobotrys* sp., radulaspores; D, *Oedocephalum* sp., radulaspores; E, *Beauveria bassiana*, sympodiospores; F, *Costantinella tillettei*, cervices and sympodiospores; G, *Exosporium tiliae*, porospores; H, *Helminthosporium velutinum*, porospores.

close together as blown out projections of the conidiophore wall. These denticles have no relationship to the growing-point of the conidiophore. When the blastospores are abstricted, the denticles may remain upon the conidiophore and give it a rasp-like appearance; for this reason such blastospores are sometimes known as **radulaspores** (from *radulum*, a rasp) (Fig. 44, C). When the blastospores are produced on sympodially proliferating conidiophores (sympodulae), they are often called **sympodiospores.**

Another type of blastospore produced not necessarily in relation to the growing-point of the conidiophore is the **porospore.** This develops as a bud extruded through a distinct pore in the wall of the conidiophore, and a pore may also be visible in the proximal end of the spore (Fig. 44, G, H). Porospores may be produced apically or laterally and in either situation may be solitary or form basipetal chains. The conidiophores are thick-walled and usually coloured, with either determinate or indeterminate branching. Porospores are sometimes produced on sympodulae.

(*b*) *Aleuriospores:* In contrast to a blastospore, an aleuriospore (syn. gangliospore; Fig. 45) is not formed by budding but instead by the apex of the conidiophore becoming inflated and delimited by a septum at an early stage. Aleuriospores thus have a wide plane of attachment, usually as wide as the conidiophore, whereas blastospores generally have a narrow point of

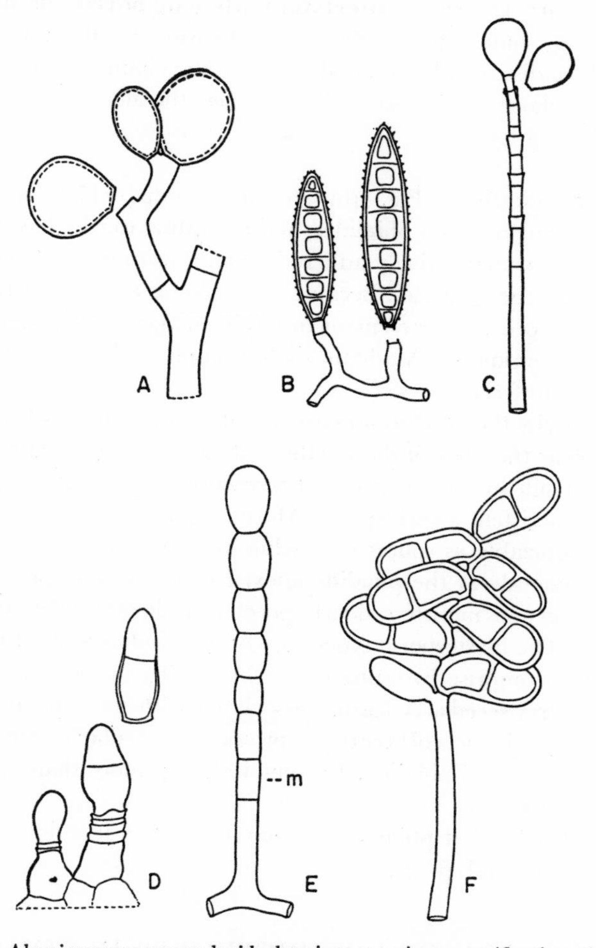

Fig. 45. Aleuriospores: camera lucida drawings at various magnifications. A, *Chalaropsis thielavioides*, with sympodial conidiophore; B, *Microsporon* sp., conidiophore with separating cell; C, *Farlowiella carmichaeliana*, annellophore; D, *Spilocaea pomi*, annellophores; E, *Oidium* sp., meristem aleuriospores with meristematic cell at *m*; F, *Trichothecium roseum*, sympodially-produced aleuriospore

attachment. Aleuriospores may be produced in basipetal succession or may be solitary; they secede by splitting of the basal septum or by circumscissile splitting of the lateral wall of a separating-cell. Those aleuriospores which are formed singly or in chains from annellophores are sometimes known as **annellospores.**

In form-genera such as *Oidium* (Fig. 45, E), the aleuriospores are formed in a basipetal chain by successive division of a meristematic cell at the base of the chain (or apex of the conidiophore). The proximal daughter cell remains meristematic, while the distal one develops into an aleuriospore. Such spores are known as **meristem aleuriospores** (or as meristem arthrospores by some authors). Because of the way in which they are formed it is difficult to distinguish the basal spores in the chain from the apical cells of the conidiophore. In the genus *Trichothecium*, the aleuriospores are formed sympodially (Fig. 45, F).

(*c*) ***Phialospores:*** these (Fig. 46) are the conidia which are abstricted in basipetal sequence from phialides, and according to whether they are dry or slimy they may respectively adhere in chains or in mucilaginous droplets at the apices of the phialides. Because a phialide is a rather distinctive type of sporogenous cell it is convenient to retain the general name phialospore, although it is now known (Madelin, 1966) that phialospores can be produced in at least four different ways.

In some species the phialospores are matured endogenously from materials inside and near the apex of the phialide; these can be termed **endophialospores** although in their manner of formation they could be regarded as endogenous meristem aleuriospores. More often, however, the phialospores become recognizable as spores only while they are in the process of being formed exogenously at the phialide apex. In some species phialospores are formed as exogenous meristem aleuriospores, while in others they are formed as blastospores. The blastosporic types may be grouped as **acrophialospores,** although they comprise two subgroups which differ in the precise means by which the spores secede. A feature associated with some phialides bearing acrophialospores is the **collarette,** a prominent cupular or recurved lip at the apex of the 'open' phialide. In some instances more than one collarette may be present.

It is noteworthy that phialospores may sometimes function as spermatia rather than as asexual spores.

AGGREGATION OF CONIDIOPHORES IN COMPOUND SPOROPHORES

Conidiophores may be borne in several different ways, sometimes as discrete simple sporophores arising from the mycelium and sometimes

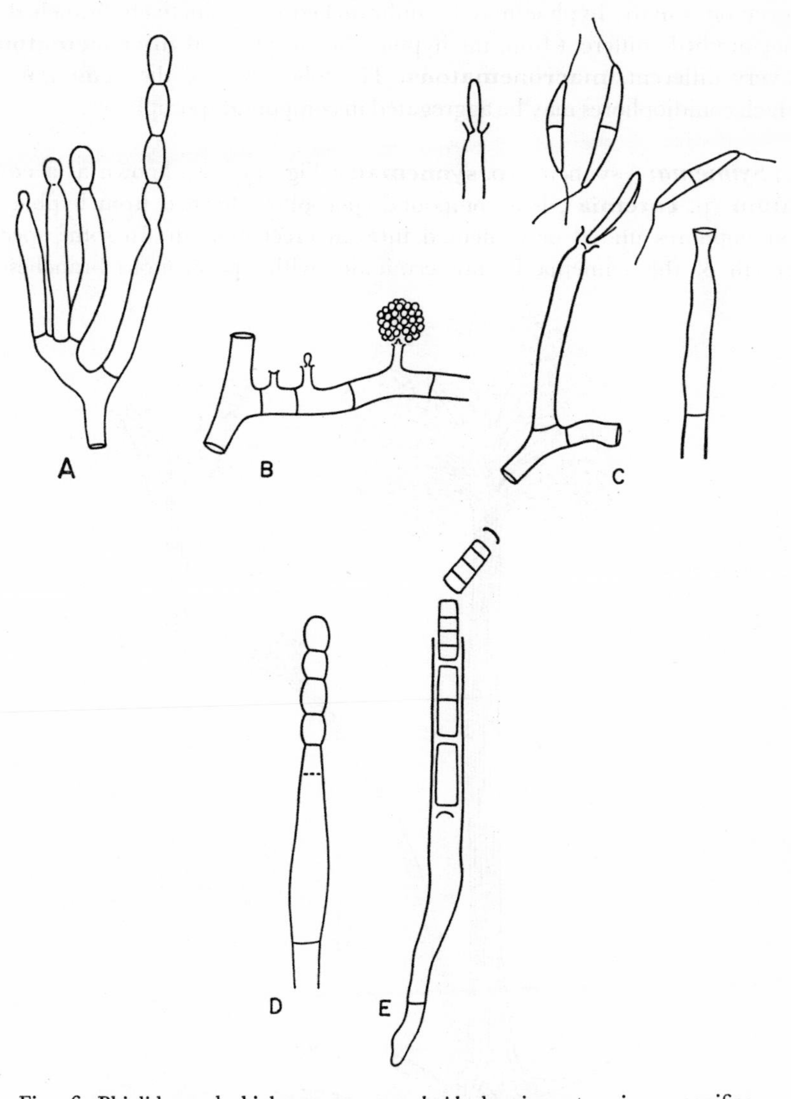

Fig. 46. Phialides and phialospores: *camera lucida* drawings at various magnifications. A, *Metarrhizium* sp., exogenous dry phialospores in basipetal chains; B, *Phialophora* sp., acrophialospores in a slimy ball; collarettes prominent; C, *Menisporopsis* sp., acrophialospores and prominent collarettes; D, a phialide producing conidia by transverse septation, conidia released by transverse splitting of walls; E, *Sporoschisma mirabile*, endophialospores. (D, after Madelin in *The Fungus Spore*, Butterworths, London, 1966.)

grouped together in compound sporophores. Conidiophores which are freely borne on somatic hyphae may be unbranched or distinctively branched. If they are little different from the hyphae they are termed **micronematous;** if very different, **macronematous.** The following are the main ways in which conidiophores may be aggregated in compound sporophores.

(a) Synnema: A synnema (p. **synnemata**; Fig. 47) also known as a **coremium** (p. **coremia**), is a compound sporophore formed from hyphae or conidiophores united or cemented into an erect column. In some species growth of the synnema is indeterminate, with sporogenous branches or

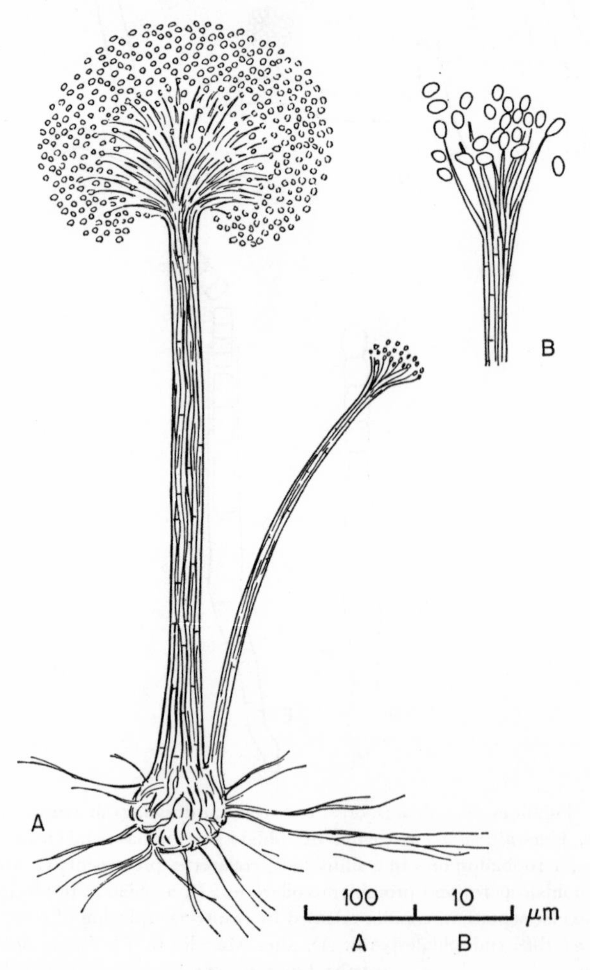

Fig. 47. Synnemata. A, B, *Graphium* sp., B, showing detail of the conidiophore apices and conidia; C, *Podosporium* sp., synnema with large phragmosporous conidia.

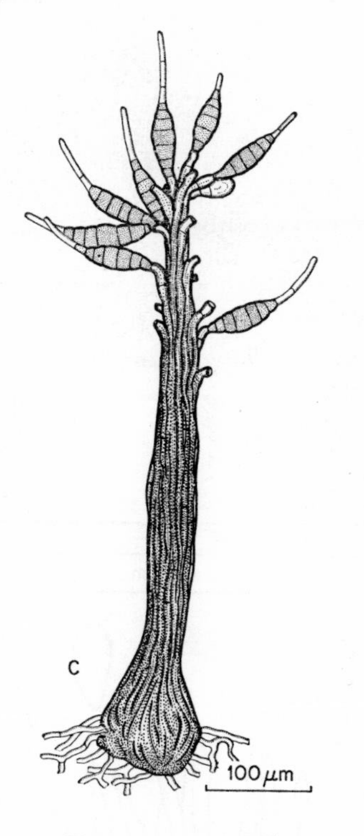

Fig. 47 — *continued.*

cells formed laterally and the apex of the synnema capable of further growth; in other species growth is determinate and the apex is the fertile part. Such differences in growth characteristics obviously have a taxonomic use in distinguishing forms producing similar sporogenous cells and conidia. The basal part of a synnema is usually sterile. Carlisle *et al.* (1961) have shown in *Penicillium claviforme* that the synnema is phototropic in the initial stages of its growth and that illumination influences its ultimate growth form; a synnema becomes much branched when it is grown in darkness before and during its period of elongation, but remains unbranched when grown in light.

(b) Sporodochium: A sporodochium (p. **sporodochia**; Fig. 48) is a compound sporophore of pulvinate (cushion-shaped) form, composed of a stromatic base giving rise to closely grouped erect conidiophores. On natural substrata the conidiophores usually erupt through bark or epidermis so that the stromatic part of the sporodochium is immersed while the fertile part is exposed. The stroma may be extensive (Fig. 48, A) or sparse (Fig. 48, B).

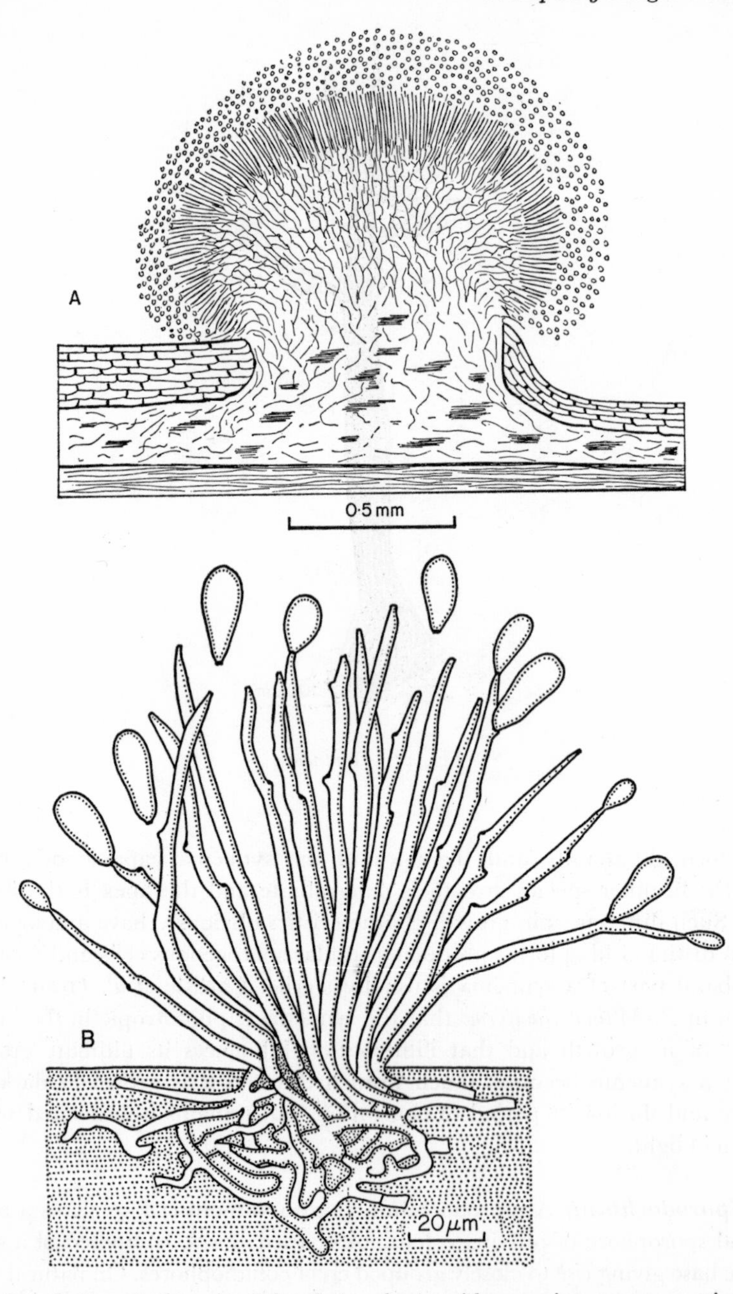

Fig. 48. Sporodochia. A, *Tubercularia vulgaris*, with extensive stroma erupting from host bark and bearing conidiophores and conidia; B, *Ramularia* sp., with loose embedded stroma and less dense sporodochium; conidiophores here are sympodulae showing scars of seceded conidia.

(c) *Acervulus:* An acervulus (p. **acervuli**; Fig. 49) is a compound sporophore composed of a basal stromatic layer and short erect conidiophores arranged in a palisade, the whole forming a flat bed covered at first by host tissues and later becoming exposed by their rupture. Some acervuli bear short or long, dark sterile hairs termed **setae** (s. **seta**), but their production is sometimes influenced by the environment; consequently their presence or absence is not a good taxonomic character, although it is used to distinguish *Colletotrichum* (with setae) from *Gloeosporium.*

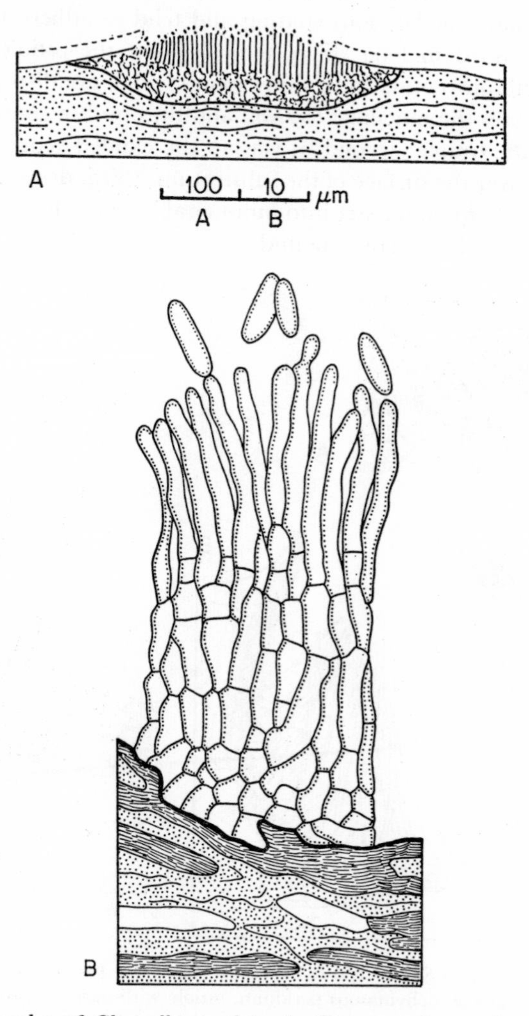

Fig. 49. Acervulus of *Glomerella cingulata.* A, Diagram of acervulus and host; B, detail showing host tissue (below) and stromatic fungal tissue bearing conidiophores and conidia (above).

(d) Pycnidium: A pycnidium (p. **pycnidia**; Fig. 50) is a globose or flask-shaped compound sporophore with a pseudoparenchymatous peridium lined on the inside by tissue giving rise to short, simple (e.g. *Phoma*) or branched (e.g. *Dendrophoma*) conidiophores usually formed in a palisade type of hymenial layer. Pycnidia may be closed, rupturing at maturity irregularly, or they may be provided with a natural pore (**ostiole**) through which the spores are liberated. The ostiole may be merely papillate or may be elongated into a 'beak', when it is said to be **rostrate.** The pycnidia may be superficial, or sunken in a stroma or in the tissues of the substratum. The conidia usually have a somewhat mucilaginous coating and tend to adhere to one another and to be extruded in long sticky tendrils from the ostiole. Most of the pycnidial Deuteromycotina have slimy spores which are dispersed by water. First, water causes the mucilage to swell and thus push a continuous mass of spores out of the ostiole; second, it washes the exuded spores and disperses them in a film over the surface of the substratum; third, drops of water falling on the film break up on impact into innumerable small droplets which carry the spores away with them on rebound.

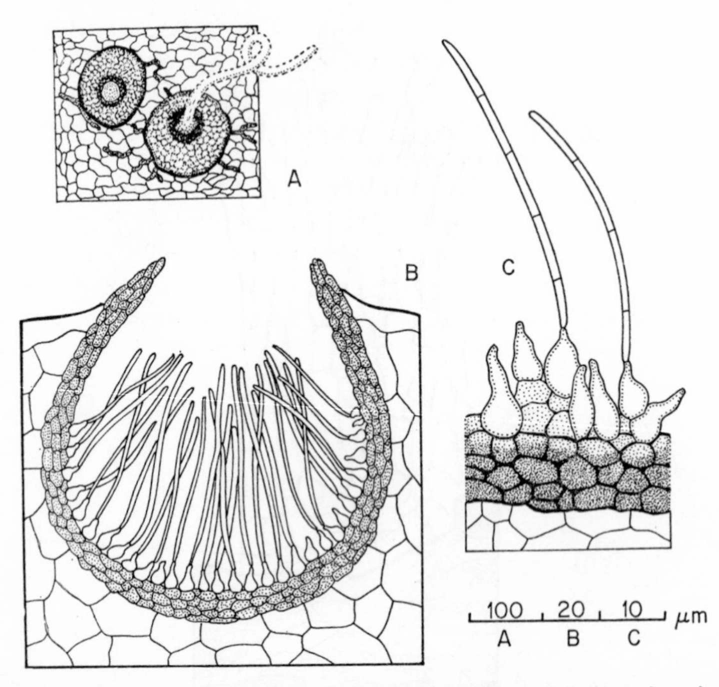

Fig. 50. Pycnidia of *Septoria apii* in celery leaves. A, two pycnidia in surface view, showing pseudoparenchymatous peridium, ostiole with dark cells surrounding it, and a cirrhus of conidia issuing from the ostiole (right); B, L.S. of pycnidium showing peridium, small flask-like conidiophores and scolecosporous conidia; C, detail of peridial cells, conidiophores and conidia.

Depending upon the substratum and environment, a single species may show considerable intergradation between some of these forms of compound sporophores which, therefore, are not entirely reliable for use in classification. Descriptions of species are usually based on materials taken from nature and occurring on natural substrata. In general the type of fructification is more constant in nature than in culture where considerable variation is often evident. But cultures often have to be used for isolation of fungi which lack characteristic natural fructifications, and in this case it may be difficult to match the fruiting structures formed in culture, with those of described species. This is particularly vexatious since many of these fungi are important pathogens.

CLASSIFICATION OF DEUTEROMYCOTINA

A natural classification of Deuteromycotina is not possible to achieve. Species with conidial states which are alike in morphology are placed in one form-genus of the Deuteromycotina, but they need not necessarily have sexual states (when these are known) sufficiently alike to occupy the same genus of Ascomycotina or Basidiomycotina. For example, species of the form-genus *Gloeosporium* have sexual states distributed among the genera *Pezicula*, *Glomerella*, *Gnomonia*, *Pseudopeziza* and *Elsinoë*. Conversely, a single genus of Ascomycotina or Basidiomycotina, comprising species with like sexual states, may contain species with a number of different conidial forms which are assigned to quite different form-genera of the Deuteromycotina; for example, asexual states of species in the genus *Ceratocystis* are placed in the form-genera *Chalara*, *Thielaviopsis*, *Chalaropsis* and *Endoconidiophora*, among others. In some instances this is certainly a reflection of the fact that the 'perfect' genera are not homogeneous in species composition and require further taxonomic examination; in others these generalizations appear to hold good, at least on present knowledge. It is clear that the genera of Deuteromycotina merely bring together morphologically related forms with similar conidiophores, conidia and conidial fructifications; but it is equally clear that these are not necessarily phylogenetically related forms. We therefore call these taxa form-genera, form-families, form-orders, etc.

The classification of Deuteromycotina still in general use is that of Saccardo (1886), presented below with slight modification (Ainsworth, 1966), but various attempts have been made by others to suggest less artificial ways of classifying these fungi.

Mason (1937) was impressed by features connected with dissemination and survival of the species and pointed out that two natural groups can be recognized, one with slime-spores (**gloiospores**) dispersed by water, insects or contact, and the other with dry spores (**xerospores**) dispersed mainly by air currents. Wakefield and Bisby (1941) adopted these distinctions to divide the Hyphomycetes into two primary groups, Gloiosporae and Xerosporae. There is no doubt, however, that the spores in a single species are sometimes slimy or dry depending upon their environment. Similarly, the type of compound sporophore found in a single species is sometimes determined by the environment. Ingold (1942) drew attention to a third biological spore-type, the aquatic spore, which is formed, liberated and dispersed under water. The majority of spores of the aquatic Hyphomycetes are large and either staurosporous or scolecosporous; spores of these shapes are supported easily by air bubbles in foam or scum, and impact more efficiently on submerged vegetation than other types of spores (Webster, 1959). A few typical aquatic Hyphomycetes are capable of occurring and sporulating outside a water habitat, while some typically terrestrial species of Hyphomycetes have been found growing and sporulating under water. Possibly, the concept of a typical aquatic Hyphomycete flora needs modifying: staurospores and scolecospores are conspicuous, whereas many of the smaller types of spores, although capable of occurring in water, are inconspicuous. Hughes (1953) showed that biological spore-types are often unstable and thus have little value in classification.

A formal classification of Deuteromycotina taking conidial ontogeny into account, has yet to be proposed. Subramanian (1962) proposed six families covering most of the Hyphomycetes of Saccardo's classification, and Sutton (1964*a*, 1964*b*) has found the principle of classification by conidial ontogeny to apply in the Sphaeropsidales. One difficulty in any attempt to classify Deuteromycotina is that a single species may produce more than one type of asexual spore, and each state would fall in a different place in the classification, also bringing about nomenclatural problems. This, of course, is also the situation with Saccardo's classification.

Saccardo's Classification of Deuteromycotina

1. No reproductive structures known — *Agonomycetes* or *Mycelia Sterilia*
   Reproduction by thallospores or conidiospores — 2.
2. Thallospores or conidiospores not borne in any form of pycnidium or acervulus — *Hyphomycetes*
   Thallospores or conidiospores borne in a pycnidium or acervulus — *Coelomycetes*

Mycelia Sterilia

One form-order, Agonomycetales; no reproductive structures are known, there being merely a sterile mycelium and perhaps sclerotia present. Examples: *Rhizoctonia*, *Sclerotium*, *Racodium*, *Ozonium*, *Rhizomorpha*.

Hyphomycetes

One form-order, Moniliales (or Hyphales); filamentous moulds.

1. Conidiophores discrete
   A. Mycelium and conidia hyaline or light-coloured — *Moniliaceae*
   B. Mycelium and/or conidia dark-coloured — *Dematiaceae*

2. Conidiophores in synnemata — *Stilbellaceae*
3. Conidiophores in sporodochia — *Tuberculariaceae*

Coelomycetes

1. Pycnidial Fungi Imperfecti — *Sphaeropsidales* (A)
2. Acervular Fungi Imperfecti — *Melanconiales* (B)

A. Sphaeropsidales

(*a*) *Sphaeropsidaceae:* pycnidia dark, leathery or membranous.
(*b*) *Zythiaceae:* pycnidia light or brightly coloured, waxy or fleshy.
(*c*) *Leptostromataceae:* pycnidia flattened, shield-shaped or elongated.
(*d*) *Excipulaceae:* pycnidia deeply cupulate when mature.

B. Melanconiales

One form-family, *Melanconiaceae*.

HOST SPECIFICITY

The Coelomycetes, particularly, contain a large number of species important as pathogens of plants, yet many of these show few distinctive morphological features which would enable them to be separated with any ease. Without more accurate means of differentiation, many species have been distinguished almost solely on the basis of the different hosts or substrata on which they were first found and described. Cross-inoculations and a study of the effects of environment on morphology would probably reveal the extent of variation and would show that many so-called 'species' are synonymous, but that they respond to different hosts or environments by showing small morphological differences. In practice, a short-cut to the identification of many Fungi Imperfecti, particularly Coelomycetes and pathogenic fungi in general, is to establish the genus by normal morphological study, then to consult a **host index** which lists the species of fungi recorded on each host or substratum. This will often eliminate all but a few species, whose descriptions can then be compared in detail with the characters of the fungus being identified.

11

FUNGI WITH ASCI AND ASCOSPORES IN FRUITBODIES: EUMYCOTA SUBDIVISION ASCOMYCOTINA

In this, the largest Subdivision of fungi, there is great morphological variety and virtually the only feature in common is the presence of asci and ascospores. The majority also form fruitbodies.

Most Ascomycotina are mycelial fungi but some, particularly the yeasts and their allies, have a unicellular or pseudomycelial thallus. The hyphae have regular primary septa with simple septal pores, occasionally with dense Woronin bodies near them but lacking a septal pore cap. No motile structures are found in Ascomycotina. The life-cycle is often pleomorphic; in many instances the perfect and imperfect states of a particular species have been connected, but there are also a large number of species known only from one or the other state (compare Deuteromycotina).

ASEXUAL REPRODUCTION

This process may occur by most of the methods already discussed: fission, fragmentation or the production of asexual spores of all types except sporangiospores. In general the asexual state is the active parasite in pathogenic species, and its spores are more important than the sexual spores in disseminating the species. The sexual state commonly forms on moribund or dead host tissues in pathogenic species, or towards the onset of winter in cold climates. In warmer places the sexual state may seldom be formed in some species and reproduction and dissemination remain predominantly asexual.

SEXUAL REPRODUCTION

Homothallic species of Ascomycotina, and also heterothallic species are found, in which compatibility is of the bipolar type. Since no motile cells are formed, planogametic conjugation is impossible; all other types of plasmogamy are possible and do occur in the Ascomycotina.

Gametangial conjugation occurs in those Ascomycotina with 'naked' asci, i.e. not formed in any kind of ascocarp. Plasmogamy between two similar gametangia, which may be simply unicellular thalli touching or coiling round one another, results in the formation of a common cell which develops into the ascus. No dicaryotic phase is present. The other types of plasmogamy occur in Ascomycotina which produce ascocarps, and the process is followed by development of dicaryotic hyphae forming tissues which eventually bear the asci.

Gametangial contact, or a reduced form of this, occurs in a large number of Ascomycotina. The dissimilar male and female gametangia, respectively called **antheridia** and **ascogonia,** are multinucleate. Transference of male nuclei into the ascogonium takes place through a pore developed at the point of contact of the gametangia, or in some species a hair-like **trichogyne** grows out of the ascogonium and makes contact with the antheridium (Fig. 51). No fertilization tube is ever formed. In species which lack antheridia, or where these are not functional, spermatia may fuse with the trichogyne or directly with the ascogonium wall. Spermatization may also occur by fusion of spermatia with somatic hyphae. In some instances plasmogamy is brought about by conidia which function as spermatia. In species which lack sex organs, plasmogamy may take place between compatible somatic hyphae, followed by the production of dicaryotic tissues from the fused cells.

Reviewing the whole range of Ascomycotina it appears that karyogamy has been progressively retarded in time and shifted in place, with the consequent production of dicaryophase tissue derived from the organs in which plasmogamy occurs. One original act of plasmogamy usually results in a considerable number of acts of karyogamy in the physiologically dependent dicaryotic tissues. Substitute types of plasmogamy, and retardation in karyogamy, are regarded by many mycologists as an indication of degeneration of sexuality; yet there is no question that they have been of biological advantage in enabling the fungi concerned to develop fruitbodies of many different types which are admirably suited to their habitats and have efficient means of liberating and dispersing their sexual spores in these habitats.

ASCOCARPS AND THEIR DEVELOPMENT

Because many asci are usually formed as the result of one act of plasmogamy they tend to be aggregated, and in the majority of Ascomycotina are formed in ascocarps. Only in relatively few (especially the Hemiascomycetes) are they unprotected. In some Ascomycotina the ascocarp tissues are differentiated around the sex organs after plasmogamy, while in others the ascocarp tissues form first, forming a stroma within which the sex organs appear later. All the tissues and structures enclosed by the peridium or other boundary layer of the ascocarp are known collectively as the **centrum** of the ascocarp; thus the centrum includes sex organs, ascogenous hyphae, asci and sterile tissues (Fig. 52). Ascocarps may occur singly or in groups, and may be superficial, erumpent (bursting through the substratum) or embedded in the substratum or host tissues. The main types of fructification are distinguished below.

1. No ascocarp: asci unprotected

Among the yeasts and most other Hemiascomycetes, an ascus is formed directly from the zygote cell resulting from gametangial conjugation. One ascus results from one act of conjugation in each case. In yeasts, the gametangia are unicellular thallus-cells and the ascus is a free cell. In *Dipodascus* the mycelial thallus differentiates multi-nucleate gametangia at the tips of certain hyphae; the gametangia conjugate in pairs and the ascus arises from the common cell formed by plasmogamy. Only one 'privileged' nucleus in each of the gametangia is functional, the remainder being supernumerary. The single diploid zygote nucleus divides by meiosis and mitosis to form innumerable haploid nuclei around which the ascospores are formed, while the supernumerary nuclei die away. It will be noted that the process just described is very like that which occurs in the Zygomycotina, except that the zygospore stage is eliminated; from this standpoint the ascus can be considered homologous with a zygosporangium.

In the Taphrinales (Fig. 56), fungi which cause leaf-curl diseases, a binucleate mycelium below the host epidermis gives rise to binucleate ascogenous cells which divide. The upper cell in each case develops directly into an ascus mother cell and then into an ascus. The asci burst through the host epidermis and, though they may be closely grouped, are unprotected by any form of peridium or stroma and thus form a hymenium of indefinite extent.

2. Cleistothecium

A cleistothecium (p. **cleistothecia**; Figs. 11, 57, 60) is a rounded, completely closed ascocarp which has no natural opening but bursts irregularly at maturity, or in a few species (e.g. *Cephalotheca savoryi*) along sutures in the peridium. Ascogonia and antheridia are formed first among the mycelial hyphae, and the tissues of the cleistothecium then form around the sex organs and the ascogenous hyphae derived from them after fertilization. In some species, the peridium is derived from mycelial hyphae and in others from hyphae originating from the base of the ascogonium. There may be considerable variation in the type of peridium enclosing the cleistothecium. In the Gymnoascaceae the peridium is composed of loosely interwoven hyphae which are intricately and characteristically branched. In the Eurotiaceae there is variation from this type of peridium to more definite forms in which the hyphae are interwoven in several layers, or become pseudoparenchymatous. The ascogenous hyphae ramify through the centrum and give rise, at different levels within the cleistothecium, to asci which are typically globose or broad, and shortly pedicellate. These usually disintegrate and release the ascospores before the cleistothecium ruptures.

In the Erysiphales the ascocarp is technically a cleistothecium because it has no natural opening, but the asci are formed in a fascicle at a single level near the base of the cleistothecium and thrust into the mass of pseudoparenchyma which at first occupies most of the centrum. These features ally the Erysiphales with those Ascomycotina which produce perithecia as their fruitbodies. The cleistothecia of Erysiphales are hibernating structures which lie dormant over winter (Ingold, 1961). Rupture of the peridium is caused by swelling of the asci as they mature, and either the asci themselves, or the ascospores, are dispersed. Sometimes the asci are thrown out violently by sudden splitting open of the tough peridium.

3. Perithecium

A perithecium (p. **perithecia**; Figs. 55, 61) is a more or less globose or flask-shaped ascocarp which opens before or at maturity by a schizogenous pore or **ostiole,** and which has its own distinct peridium of specialized tissue enclosing the centrum.

In species where functional antheridia and ascogonia occur, or where antheridia are suppressed and the ascogonia are spermatized, the sterile tissues forming the perithecial wall are usually formed from hyphae which branch from the stalk cells below the ascogonium, and the centrum tissues develop from branches of the ascogonium itself. Thus the centrum and the peridium build up almost simultaneously. But most species lack sex organs, and the picture is somewhat different. Here, the perithecia are often formed from special coiled hyphae which are thought to undergo spermatization or somatogamy and then give rise to ensheathing branches which form a plectenchymatous mass, or stroma. It is not clear whether the ascogenous hyphae which subsequently form within the stroma have their origin in the coiled hyphae, though this would seem likely; at all events, the ascogenous hyphae and the asci developed from them thrust their way into the stromatic tissue, most of which is lysed or crushed leaving a few layers of peridial pseudoparenchyma towards the outside of the perithecium.

Most perithecia have true **paraphyses** which arise from the innermost layer of peridial tissue and project as sterile branches between the asci. The paraphyses may be simple or branched, sometimes coloured, sometimes inflated at the apex (**capitate**), or at intervals along their length (**moniliform**), and with or without close septation. Paraphyses are said to space out the asci and to secrete mucilage which is concerned in spore discharge mechanisms; but space is also made by centrifugal expansion of the growing peridium. Sometimes the ostiole is prolonged into a beak (**rostrum**) and usually it is lined with short hair-like **periphyses** (s. **periphysis**) directed towards the opening and assisting in expelling asci or ascospores. Typically, the hymenium lines the inside of the perithecium as a palisade of asci, or it may take the form of a basal fascicle of asci formed at a common level; the asci are not scattered at different levels within a perithecium.

True perithecia of the type described may sometimes occur embedded in a stroma composed of hyphal tissue alone, or of hyphal and host tissues combined. The stroma may not necessarily surround the perithecia, but may instead be reduced to a shield-like **clypeus** around the ostiole, or to a **subiculum** of stromatic interwoven hyphae below the perithecium.

The liberation and discharge mechanisms of perithecia have been discussed by Ingold (1965). Ascomycotina with perithecia (the Pyrenomycetes) are generally xerophytic and the perithecia have a firm peridium and often a positively phototropic ostiole. When the necessary water is present, the centrum tissues become fully turgid. In some types (e.g. *Chaetomium*) ascospores are liberated non-explosively: the ascus walls break down and the ascospores exude through the ostiole in a mass of mucilage which swells as it absorbs water and can then no longer be contained in the perithecium. In some of the types with long rostrate perithecia, the asci themselves become detached and absorb water: the resulting pressure within the perithecium forces the asci out, one by one, through the narrow rostrum and they may burst and liberate their spores as they pass through the ostiole. With most Pyrenomycetes one ascus only matures and discharges at a time. It elongates by its own growth and by pressure from other asci in the perithecium, until its apex projects just beyond the ostiole, where it bursts as a result of turgor pressure and liberates the ascospores simultaneously or in quick succession. The ostiole is often too narrow to permit more than one ascus to elongate through it at one time.

4. Ascostroma

Many Ascomycotina form their asci in cavities (**ascolocules**) in a stromatic fruitbody known as an **ascostroma** (p. **ascostromata;** Fig. 69). This is composed of somatic hyphae modified into pseudoparenchyma and is therefore not unlike a sclerotium before the asci start to appear. In some ascostromatic groups (e.g. Dothideales), the stroma develops before ascogonia are formed within it, and the ascolocules result from crushing or disintegrating of the pseudoparenchyma by the developing ascogenous hyphae and asci. In other groups (e.g. Pleosporales) the ascolocules appear in the stroma as it expands and before the asci are formed. With either method of formation an ascolocule lacks its own distinctive peridium, being bounded simply by the pseudoparenchyma of the stroma; but where there is only one ascolocule in a much reduced stroma, it may be difficult to distinguish it from a perithecium, in which case various associated features become important differential aids. In most instances the presence of an ascostroma is correlated with bitunicate asci, and a perithecium with unitunicate asci (to be discussed below); also an ascolocule does not contain paraphyses or periphyses, but may contain pseudoparenchyma or pseudoparaphyses. **Pseudoparaphyses** (Fig. 69) are sterile thread-like hyphae attached to both the roof and the base of the ascolocule, and are formed by separation of hyphae during the expansion of the ascostroma, as in the Pleosporales.

If the ascostroma is reduced to one locule with or without pseudoparaphyses, and resembles a perithecium, the fruitbody is known as a **pseudothecium.** Such an ascostroma usually opens by means of a lysigenous pore or slit which is known as an ostiole although its method of development is unlike that of a perithecial ostiole. An interesting type of ascostroma, in appearance intermediate between an apothecium and a perithecium, is the **hysterothecium,** characteristic of the Hysteriales. This is an elongated linear, carbonous ascostroma with a long cleft-like opening parallel to the long axis (Fig. 14, D, E); the centrum structure and bitunicate asci show that this type of ascocarp belongs to the ascostromatic group. Another distinctive type of ascostroma characterizes the Microthyriales or Hemisphaeriales; this is the **uniloculate thyriothecium** (Fig. 69), a small flattened ascostroma like an inverted saucer or shallow volcano, whose uppermost surface is typically radiate in construction and is ostiolate.

Since the asci are embedded in, or surrounded by, pseudoparenchyma, they can discharge their spores only when an ostiole has been lysed in the tissue or when the pseudoparenchyma has absorbed water, undergone gelatinization, and then broken down.

5. *Apothecium*

An apothecium (p. **apothecia**; Figs, 51, 52) is an open ascocarp bearing the asci in an exposed hymenium. Typically, the apothecium is discoid or cupular but several modifications of this form are encountered. The most important of these are, incurving of the hymenium to form deeply cupulate apothecia which eventually (in types formed below ground; Tuberales) are replaced by closed, solid fructifications; or recurving of the hymenium to form clavate or spathulate fruitbodies with the hymenium spread over the whole outer layer or confined to an upper fertile part called the **pileus,** borne on a stipe. Apothecia may, however, be either sessile or stipitate. They may also be compounded so that a number of shallow fertile apothecial pits

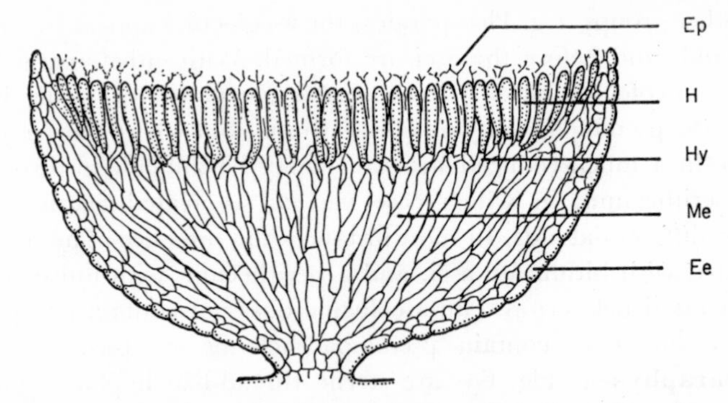

Fig. 51. Hypothetical apothecium drawn diagrammatically to show tissue distribution. *Ee*, ectal excipulum; *Ep*, epithecium; *H*, hymenium; *Hy*, hypothecium; *Me*, medullary excipulum.

or depressions, separated by sterile plates of hyphal tissue, are formed towards the outside of a common stroma.

Paraphyses may be present or absent in the hymenium of apothecia of different species; more commonly they are present. They usually terminate at about the level of the ascal apices, but sometimes project above the asci and there form a layer termed the **epithecium.** The outer boundary layer of wall tissue surrounding the apothecium is known as the **ectal excipulum** and the inner part as the **medullary excipulum,** while the layer of tissue immediately beneath the hymenium is called the **hypothecium** (Fig. 51).

Apothecia may be **angiocarpic** in development, the hymenium being formed within an enveloping mass of hyphae to begin with, or **gymnocarpic** if the hymenium is superficial from the outset. A middle course of development, **hemiangiocarpic,** is known in some species (Corner, 1929). However, in all forms of apothecia the hymenium is exposed at maturity, with the exception of the hypogean fruitbodies of the Tuberales.

The classical species studied in connexion with plasmogamy and subsequent formation of an apothecium is *Pyronema omphalodes* (=*P. confluens*) which occurs on burnt or steam-treated soil, or charcoal (Fig. 52). Pairs or groups of ascogonia and antheridia appear in certain places on the mycelium. Each of these is multinucleate, with haploid nuclei, and is derived from a hypha of similar nuclear constitution. The ascogonium has a few stalk-cells at its base and a trichogyne which makes contact with the antheridium. Compatible male and female nuclei pair off in the ascogonium but do not immediately fuse; they migrate as dicaryons into ascogenous hyphae which grow out from the wall of the ascogonium. Thus the ascogenous hyphae are dicaryotic. At the same time, hyphae branch from the ascogonial stalk-cells to form the sterile haplophase paraphyses which are interspersed between the asci in the mature fructification. The hyphae forming the excipular

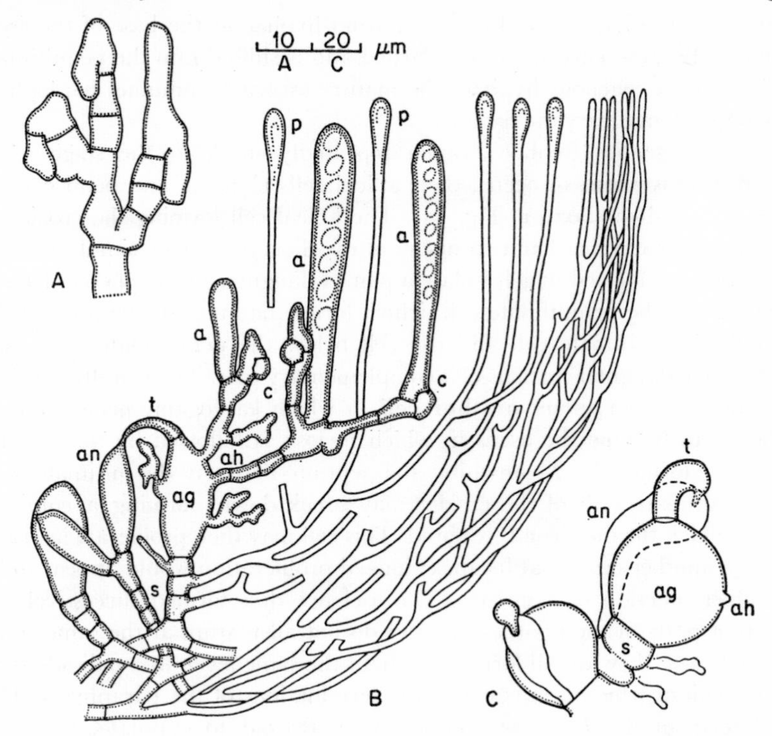

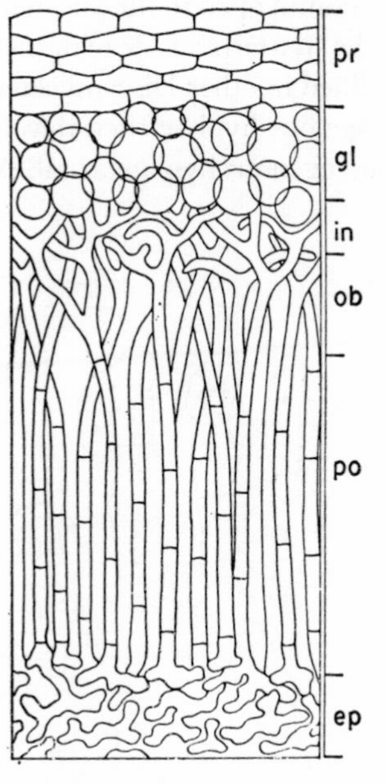

Fig. 52. A, Croziers in *Trichophaea* sp. showing both apical and lateral proliferation; B, a hypothetical apothecium showing development of ascogenous hyphae, croziers and asci from the fertilized ascogonium, and paraphyses and excipular tissue from sterile cells below the ascogonium; C, ascogonia and antheridia of *Pyronema omphalodes*; *a*, ascus; *ag*, ascogonium; *ah*, ascogenous hyphae; *an*, antheridium; *c*, crozier; *p*, paraphysis; *s*, stalk-cell; *t*, trichogyne; D, highly differentiated tissue of a Discomycete fruitbody; [Diagrammatic after Starbäck, Bihang till K. Sv. Vet.-Akad. Handl. 1895, **21** (iii), 5.)] *ep*, textura epidermoidea; *gl*, tex. globulosa; *in*, tex. intricata; *pr*, tex. prismatica; *po*, tex. porrecta; *ob*, tex. oblita.

layers are also haploid and originate from hyphae at the base of the ascogonium. Because karyogamy is retarded and is shifted into the penultimate cells of the ascogenous hyphae, the mature ascocarp combines haplophase and dependent dicaryotic tissues.

The ascogenous hyphae branch repeatedly until, at the stage when karyogamy is about to occur, their apical cells elongate and bend over to form a hook-like **crozier** (Fig. 53). The apical cell forming the crozier is, of course, dicaryotic. The two nuclei of the dicaryon divide simultaneously and septa are formed which isolate a pair of daughter nuclei of complementary type in the penultimate cell, which forms the bulge at the apex of the crozier. The ultimate cell and the cell below the penultimate one, each receive one daughter nucleus of a complementary pair. The penultimate cell now becomes the **ascus mother cell** in which karyogamy occurs, and it enlarges to become the ascus in which meiosis is followed by mitosis and ascospore-formation. Meanwhile, the septum between the ultimate and subpenultimate cells of the crozier may break down, forming a common cell in which the dicaryon is restored. This cell may then proliferate laterally to form another crozier at its apex; indeed proliferation would appear to be the chief function of a crozier. The asci are thus formed successively in different parts of the apothecium and do not all mature at the same time. Nevertheless, they are all formed at about the same level, in a palisade type of hymenium which also contains the terminations of the paraphyses. The final form of the *Pyronema* apothecium is discoid to cupulate, with the hymenium open to the air on its upper, or superior, surface.

The pattern of development shown in *Pyronema* may be considerably modified in other Discomycetes. In some species there are no croziers and the ascogenous hyphae then develop directly into asci. In others, there may be no ascogenous hyphae at all. Ascogonia may lack trichogynes and conjugate

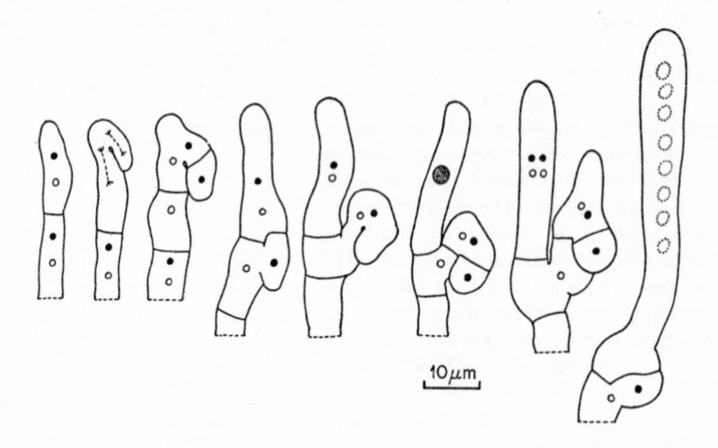

Fig. 53. Crozier formation in *Peziza* sp. (only the cytology is diagrammatic).

directly with antheridia, or the ascogonia themselves may lose their morphological identity and become more like somatic hyphae. The greatest changes appear to be attributable to functional or complete degeneration of the antheridia; in these cases the ascogonial nuclei may fuse parthenogenetically, or may be fertilized by spermatization. Finally, no sex organs may be formed and somatogamy becomes the only possible method of plasmogamy.

Liberation and discharge mechanisms in apothecial Ascomycotina (Discomycetes) are, as Ingold (1965) has shown, beautifully correlated with other biological features. First, cupulate apothecia face upward and can hold water, thus their liberation mechanism is usually one of 'water-squirt' by bursting of the turgid asci. As the ascospores are explosively discharged to a distance up to 4 cm, they are readily dispersed in air currents. The stipitate-pileate types of apothecia also have their hymenia exposed to water and are sufficiently tall for the spores to be dispersed without falling back upon the substratum. Second, the asci are usually positively phototropic and either curve as a whole or have their apices or opercula displaced towards the light source. In a deeply cupulate apothecium this has the advantage that the spores of asci which line the sides of the apothecium are not discharged towards the opposite side, but instead towards the opening of the apothecium. It is also advantageous in discoid apothecia occurring on soil, dung or litter — situations where pieces of the substratum might intercept spores if they were not discharged towards light. In some Discomycetes (e.g. *Ascobolus*) relatively few asci mature and project above the hymenial level each day. But in many of the large Discomycetes (e.g. *Peziza*) thousands of asci in the hymenium mature each day and reach a state of unstable equilibrium so that the least change in humidity, temperature, air turbulence or light intensity is enough to cause simultaneous bursting of the asci and discharge of their spores. Sufficient numbers are involved for there to be an audible 'hiss' as the asci burst and a visible 'puff' of spores. Where the ascospores are scolecosporous or cylindrical (e.g. *Trichoglossum*; Fig. 67) only one spore is discharged at a time; it plugs the pore of the ascus until sufficient turgidity is built up to eject it, and is then replaced by another spore in the series until finally, all are discharged and the empty ascus collapses.

ASCI AND ASCOSPORES

Asci may be sessile or pedicellate. Most are elongated, cylindrical or clavate, but globose or pear-shaped (**pyriform**) asci are more common in those groups in which a dense palisade hymenium is not formed, possibly because in a discontinuous hymenium there is little resistance to lateral expansion of the asci. There is also a tendency for asci to be rounded when they are surrounded by stromatic tissue in which, presumably, the formative pressures are equal on all sides.

Two general types of asci may be distinguished on the basis of wall structure: bitunicate and unitunicate asci.

Bitunicate asci

A bitunicate ascus (Fig. 54) is one with two distinct, separable walls. In fresh material mounted in water the outer wall (**ectoascus**), which is thick and inextensible, ruptures transversely near the apex to permit the thin inner wall (**endoascus**) to expand into a long cylindrical sac with an apical pore through which the ascospores are discharged in succession. The apex of the ectoascus may be forced off as a thimble-like cap, while the basal part often forms a constricted, wrinkled collar around the lower part of the endoascus.

When rupture and dehiscence of the ascus cannot be observed, one has to resort to associated characters for diagnosing a bitunicate ascus (Dennis, 1960). The apex is usually very thick and often indented on its inner-wall surface; it never colours blue with Melzer's chloral-hydrate iodine solution — i.e. it is **non-amyloid** — and it lacks an operculum or a plugged pore or peculiar ring-like thickenings which are common in unitunicate asci. The asci themselves tend to have short, sharply delimited stalks. The ascospores are frequently, but not invariably, coloured and multicellular. Again, with exceptions, the ascocarps tend to be very small and closed. Bitunicate asci are typical of the Loculoascomycetes.

Unitunicate asci

A unitunicate ascus has a single wall, even when this is laminated in structure, and the wall is either uniformly thin, or thickened at the apex. Two classes of unitunicate asci are distinguished by their respective means of dehiscence. **Operculate** asci have an **operculum** (Fig. 54) at the apex: a hinged lid-like opening. The remainder of unitunicate asci are **inoperculate,** which means that they may be either indehiscent, with no natural opening, or may dehisce through an apical pore or slit. The indehiscent asci liberate their ascospores either by deliquescing or by becoming turgid and bursting; the dehiscent types, both operculate and inoperculate, become turgid and discharge their ascospores forcibly, simultaneously or in quick succession.

In dried material it may be difficult to establish whether an ascus is operculate or inoperculate and again, reliance must be placed on associated characters; these have been usefully summarized by Dennis (1960). The test for amyloidity of the ascus tip is not always conclusive since a positive or a negative reaction may be obtained with both operculate and inoperculate asci in different groups of fungi. However, if the apex is thick and penetrated by a canal which at first is blocked by a plug, it can be assumed to be inoperculate; and in most species this plug is amyloid and thus readily detected. Again, the ascus is inoperculate if its apex has a thickened ring on the

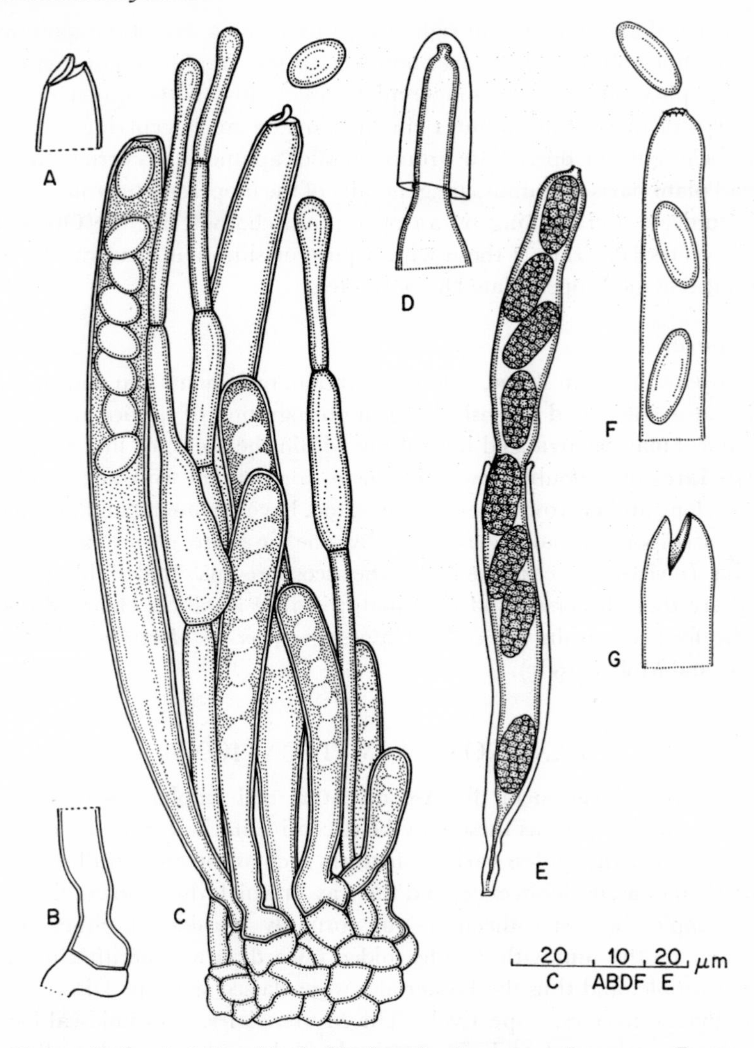

Fig. 54. *Peziza domiciliana* (A, B, C). A, unitunicate operculate ascus; B, ascus base showing remnant of crozier; C, asci and paraphyses; D, *Tympanis alnea*, apex of bitunicate ascus with ectoascus as a cap; E, *Pleospora herbarum*, bitunicate ascus with expanded endoascus projecting above thick ectoascus; F, *Dasyscypha wilkommii*, inoperculate ascus dehiscing by a pore; G (diagrammatic), inoperculate ascus dehiscing by a slit.

inner wall surface, represented in optical section by two small dots on opposite sides of the ascus; in many species this ring is amyloid. The ascospores of operculate asci are never septate, are often ornamented, and are either symmetrical or asymmetrical about the long axis. In inoperculate asci the ascospores are always smooth, either septate or scolecosporous, or asymmetrical about an axis at right angles to the long axis of the spore when they are not septate. There are, however, a few troublesome genera with smooth, spherical spores in inoperculate asci. Most of the larger cupulate ascocarps of more than about 1 cm in diameter and occurring on soil or dung, belong to the operculate group; smaller apothecia occurring on living or dead plant parts or humus, are generally of the inoperculate group.

Unitunicate asci opening by an operculum characterize the Operculate Discomycetes (Pezizales); those with a pore or slit, or no natural opening, are typical of the Inoperculate Discomycetes.

Ascospores

Ascospores differ in shape, colour, septation, ornamentation and size, and these features are used extensively in the recognition of species and genera. The spores may be arranged irregularly within the ascus, or in a single row (**uniseriate**) or a double row (**biseriate**), or parallel with one another if they are long and narrow. These features, too, have taxonomic significance.

The ascospores usually germinate by one or more germtubes to form mycelia. In yeasts or yeast-like forms, the ascospores may bud off blastospores which are then dispersed and germinate; in rare instances chlamydospores may be formed within the cells of ascospores, as in *Mycosphaerella pinodes* (Carter and Moller, 1961).

TAXONOMIC IMPLICATIONS

Modern classification of the Ascomycotina still emphasizes the various types of ascocarps but, as these may be deceptive in their gross appearance, it also relies heavily on less variable features such as the ascus wall structure, the method of ascus dehiscence, and the way in which the centrum develops. For example, a hysterothecium was formerly considered either as an elongated apothecium with its sides folded inward, or as a perithecium with a linear ostiole, and thus the Hysteriales were classed with the Discomycetes or the Pyrenomycetes, respectively. The Myriangiales, with uniascal locules in a stroma, appeared to have perithecia and were regarded as Pyrenomycetes. However, the centrum-development and bitunicate asci of both Hysteriales and Myriangiales show that these orders are best classed in the Loculoascomycetes. Recognition of these features has facilitated the classification of several difficult groups and has placed them in better relationship with one another. (Nannfeldt, 1932; Luttrell, 1951, 1955, 1965; Muller and von Arx, 1962; von Arx and Muller, 1954; Dennis, 1960.)

Classes of Ascomycotina

The chief distinguishing features of the classes of Ascomycotina are shown in the following key. The classification adopted is a compromise between older systems in which the nature of the ascocarp was emphasized to the virtual exclusion of other characters, and newer ones in which microscopic characters are more prominent.

1. Asci unitunicate, or if bitunicate then in an exposed hymenium of an apothecium — 2.
   Asci bitunicate, formed in an ascostroma but not in an apothecium — 5.
2. Asci naked, i.e. formed as discrete free cells or in a hymenium of indefinite extent, not bounded by a stroma or by ascocarp tissue; asci indehiscent — *Hemiascomycetes*
   Asci formed in ascocarps — 3.
3. Asci scattered at various levels within a cleistothecium or a beaked perithecium; asci indehiscent — *Plectomycetes*
   Asci forming a hymenium or arising as a fascicle at a common level in the ascocarp, or rarely single — 4.
4. Ascocarp usually a perithecium, less often a cleistothecium with fasciculate asci or an ascostroma with unitunicate asci; asci inoperculate, with an apical pore or slit; not minute external parasites of insects and arachnida — *Pyrenomycetes*
   Ascocarp a perithecium with inoperculate asci whose walls soon disintegrate; minute external parasites of insects and arachnida — *Laboulbeniomycetes* (Fig. 55; not considered further)
   Ascocarp an apothecium or its hypogean derivative; asci operculate, inoperculate or indehiscent — *Discomycetes*
5. Asci bitunicate, formed in an ascostroma but not in an apothecium — *Loculoascomycetes*

CLASS HEMIASCOMYCETES

This class comprises those Ascomycotina which form unitunicate asci as free cells or in a hymenium of indefinite extent. The yeasts and their mycelium-forming or pseudomycelial relatives constitute the order Endomycetales, while the pathogenic leaf-curl fungi are placed in the order Taphrinales. In the Endomycetales the asci arise from a common cell formed by gametangial conjugation; in the Taphrinales the binucleate mycelium gives rise to binucleate ascogenous cells which develop into asci. Paraphyses are not formed by members of either order.

Order Endomycetales

Four families are usually included in this order. The Ascoideaceae includes genera with multisporous asci borne on a filamentous mycelium, e.g. *Ascoidea* and *Dipodascus*, both of which have been isolated from gummy exudates of trees. In the Spermophthoraceae, a mycelium is also present but the asci have eight or fewer ascospores and these are fusoid or narrow and elongated; members of this family are pathogens of tropical plants, especially cotton, and are introduced to their hosts by insect punctures. The Endomycetaceae are distinguished from the Spermophthoraceae by producing

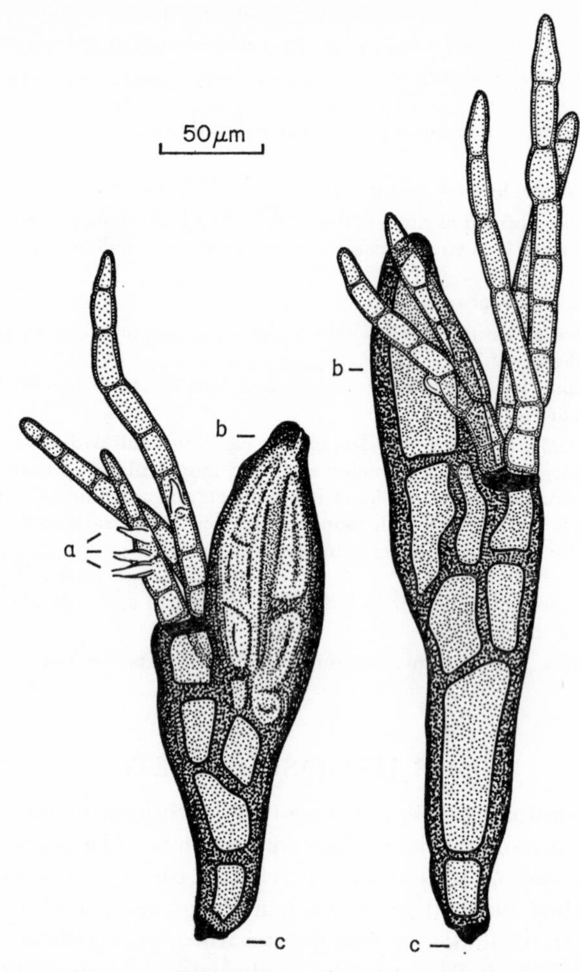

Fig. 55. *Laboulbenia elongata*, two thalli taken from the elytra of *Agonum marginatum*; *a*, antheridium; *b*, perithecium; *c*, attachment.

uninucleate instead of multinucleate gametangia and by their occurrence mainly on sugary substrata. The fourth family, Saccharomycetaceae, will be considered in a little more detail.

Family Saccharomycetaceae

Included in this family are the members of the Endomycetales with single-celled or pseudomycelial thalli and which form eight or fewer ascospores per ascus. These, then, are the commercial yeasts and related forms, important because of their fermentative powers and because certain species are able to cause disease in man and animals. Species of *Saccharomyces* are used to ferment sugar with the production of carbon dioxide (in baking) or alcohol (in brewing), but yeasts may also ferment foodstuffs and spoil them. Yeasts are also a valuable source of vitamins. *Candida albicans* causes the diseases known as thrush and sprue in the mucosa of man and animals; the presence of chlamydospores in its pseudomycelium is a useful aid in distinguishing this from similar, but non-pathogenic, yeasts. *Histoplasma capsulatum* and *Cryptococcus neoformans* are dimorphic species with yeast-like states in their life-cycles, and respectively cause serious diseases of the lymphatic system, and the lungs and meninges, in man.

In general a yeast thallus is unicellular, but the cells may often be united in chains formed by incomplete budding or fission, thus producing a pseudomycelium. On agar media the colonies formed by yeast cells are moist, glutinous and often wrinkled; in liquid media they tend to be ropy. Even in a single species there may be much variation in the size and shape of the cells, while their relatively simple morphology makes classification difficult. Thus the classification of yeasts is based on whether asexual reproduction occurs by budding or by fission, on whether ascospores are produced, and then largely on physiological and biochemical differences. In species which produce ascospores, further distinctions are to be found in the number of ascospores per ascus, and in their shapes. While the majority of yeast ascospores are oval or subspherical, some are shaped like a bowler-hat (*Pichia*) or like the planet Saturn (*Hansenula*). In *Kluyveromyces polysporus* the asci are exceptional in being multisporous, thus suggesting the type of ascus found in the Ascoideaceae.

Three types of life-cycle are known in yeasts, differing mainly in the preponderance of either the haplophase or the diplophase. In *Saccharomyces cerevisiae*, the common baker's yeast, both phases last a long time and are carried forward by bud-cells.

Order Taphrinales

Species of *Taphrina* are responsible for leaf-curl diseases of many plants, particularly stone-fruits of the genus *Prunus*. *Taphrina deformans* and *T. aurea* cause leaf-curl of peach and poplar, respectively. All species are pathogenic, causing malformations such as curling, puckering or blistering. The effects may be rather similar to hypertrophy caused by mites.

In culture, no mycelium is formed and the colonies are yeast-like. In nature, a fairly extensive mycelium is present in the host leaves. Binucleate cells below the host epidermis become differentiated as ascogenous cells without any form of plasmogamy occurring. In *Taphrina deformans* (Fig. 56) the ascogenous cell elongates after karyogamy and becomes divided transversely by a septum into a stalk-cell and an upper ascus mother cell at the same time as the diploid nucleus divides by meiosis. In *Taphrina aurea* there does not appear to be a stalk cell (Fig. 56). The asci in all species erupt

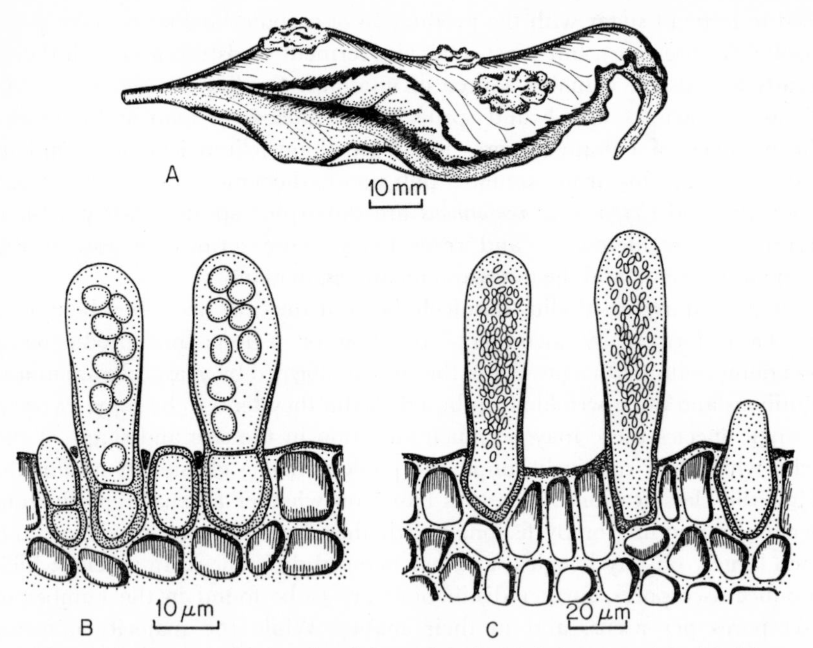

Fig. 56. A, Hypertrophied and curled peach leaf with lesions caused by *Taphrina deformans*; B, *Taphrina deformans*, stalk cells and asci with ascospores; C, *Taphrina aurea*, asci filled with conidia budded from the ascospores.

through the host epidermis, giving it a waxy appearance; they contain eight ascospores which frequently bud off innumerable blastospores while still within the ascus. The ascospores and blastospores are uninucleate and haploid; during germination, their nuclei divide mitotically to initiate the binucleate mycelium which infects the host. The cells are therefore not dicaryotic but instead contain pairs of sister nuclei.

CLASS PLECTOMYCETES

Typically, members of the Plectomycetes have cleistothecial ascocarps, lacking paraphyses but containing ascogenous hyphae and asci formed at various levels. The asci are globose or broad, and indehiscent; they deliquesce and thus liberate the ascospores within the ascocarp. It is these centrum characteristics rather than the actual form of the ascocarp that are considered taxonomically most important, and for this reason the orders Microascales and Onygenales are included as Plectomycetes. The Onygenales are of interest because some of their members form small stipitate fructifications on horn, hooves, feathers and hair of dead animals, substances which are alike in being composed of keratin. The asci in *Onygena* are globose and soon dissolve away to leave a dry powdery mass of ascospores interspersed with a hyphal capillitium. In the Microascales the ascocarp is a beaked perithecium, usually partly sunken in the woody substratum, but the asci are broad or rounded cells which soon dissolve and leave the ascospores free to be discharged in mucilage, which gathers as an opalescent droplet at the opening of the ostiole. The best known genus of Microascales is *Ceratocystis*, some of whose species cause notable diseases of hardwood trees, e.g. Dutch Elm disease (*Ceratocystis ulmi*) and Oak wilt disease (*Ceratocystis fagacearum*). Coniferous wood is rendered unsightly by 'blue-stain' caused by *C. pilifera*, or by other species such as *C. ips* which are distributed by bark-beetles and provide a mycelium in the larval tunnels for the larvae to feed on (compare mycangia, p. 68). Much variety exists in the imperfect states of different species of *Ceratocystis*. Some species produce moist exogenous phialospores in a droplet at the apex of a simple phialide; others form dry endogenous phialospores which adhere in chains; in still others the conidiophores form a synnema which bears a mucilaginous drop of conidia at its apex.

The third order of Plectomycetes, the Eurotiales, will now be dealt with in some detail.

Order Eurotiales

Members of the Eurotiales produce cleistothecia on the mycelium or within stromatic tissue having the appearance of a sclerotium. Here, we shall be concerned only with the family Eurotiaceae and in particular the form-genera *Aspergillus* and *Penicillium*. Nevertheless, some genera in other families are of interest. *Ascosphaera* species (Ascosphaeraceae) inhabit beehives, where they cause disease of the bee larvae and decay of pollen in the honeycombs; the asci in this genus are united into compact groups ('spore-balls', or more correctly ascus-balls) and are formed within the ascogonium. Many of the dermatophytes — i.e. fungi causing diseases of the skin and hair in man and animals — have perfect states which belong in the Gymnoascaceae; these are keratinophilic and are also found saprobically on leather, feathers, hair and other such substances. The Amorphothecaceae are represented by *Amorphotheca resinae* (Parbery, 1969), a species whose imperfect state, *Cladosporium resinae*, is important for its occurrence in creosote, diesel oil and aviation kerosene, where its mycelium is obviously a danger because it may block fuel systems. The perfect state of this mould is of particular interest because its cleistothecia have peridia composed of a dark amorphous substance secreted around the developing centrum, and not of hyphae or their derivative tissues.

Family Eurotiaceae

In dealing with the Eurotiaceae we are faced with a nomenclatural problem. Many of the Eurotiaceae are best known, or known only, in their asexual states comprising the form-genera *Aspergillus* and *Penicillium*, which are ubiquitous and highly important moulds from many points of view. Different species of *Aspergillus* may have perfect states which differ in their methods of cleistothecial development, or in centrum characteristics, and are accordingly placed by some authors in at least four perfect-state genera: *Eurotium, Emericella, Sartorya* and *Hemicarpenteles*. A similar situation in *Penicillium* species has led to the recognition of two perfect-state genera, *Talaromyces* and *Eupenicillium*. Other types of development are known in a few other species, but as yet no new perfect-state genera have been proposed for them. Until more is known about the perfect states, it would be convenient to follow the eminent monographers of the Aspergilli and Penicillia (Thom; Raper; Fennell) who have recommended that the names *Aspergillus* and *Penicillium* should be used for both asexual and sexual states of these fungi. But as this proposal is a violation of the Code of Nomenclature we shall circumvent the issue in the following account, by approaching the Eurotiaceae through a consideration of their imperfect states.

***Form-genus* Aspergillus:** The Aspergilli (Fig. 57) are very common moulds living on a vast number of different substrates, presumably because they are able to produce a variety of different enzymes. Various species of *Aspergillus* are concerned in decaying foodstuffs, leather and textiles, causing lung disease (aspergillosis) of birds and humans, and in causing infections of the ears and tongue in man. Some species are used in fermentations to produce citric and gluconic acids, and certain antibiotics. *Aspergillus flavus* may produce an extremely potent toxin, aflatoxin, in stored grain and fodder fed to farm animals. The amount of toxin formed depends partly upon the particular strain of *A. flavus* present, but can also vary greatly when a single strain is grown on different media, i.e. toxin-production is controlled partly genetically and partly physiologically. *Aspergillus niger*, a very common laboratory contaminant, can be used to detect minute traces of copper, the colour of the spores of some strains varying with the amount of copper present.

In *Aspergillus*, the hyphae are branched and septate, with multinucleate cells. A conidiophore is formed as a vertical branch of a horizontal cell which remains evident as a 'foot-cell', and the conidiophore ends in a multinucleate vesicle. Phialides are produced in one rank directly from the vesicle, covering its entire surface, or in a secondary rank as branches of **metulae** (s. **metula**) which arise from the vesicle. The conidia are formed in basipetal chains from the phialides; they are dry, dusty, typically spherical and ornamented with small spines; they are dispersed by air currents. They vary in colour from almost colourless to green or blackish in different species.

I have already indicated that at least four perfect-state genera are associated with the asexual states of *Aspergillus*. In *Eurotium*, antheridia, ascogonia and ascogenous hyphae are present and the cleistothecium builds

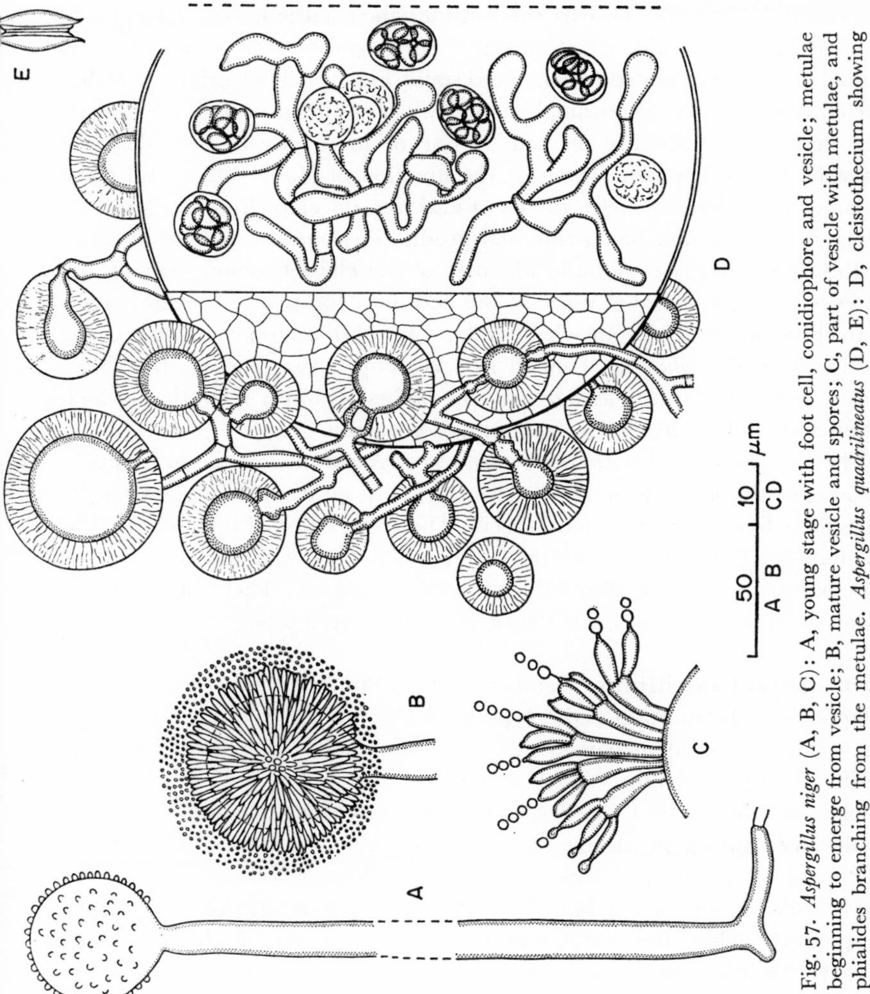

Fig. 57. *Aspergillus niger* (A, B, C): A, young stage with foot cell, conidiophore and vesicle; metulae beginning to emerge from vesicle; B, mature vesicle and spores; C, part of vesicle with metulae, and phialides branching from the metulae. *Aspergillus quadrilineatus* (D, E): D, cleistothecium showing towards left the Hülle cells enveloping the peridium; right, the interior with scattered ascogenous hyphae, asci and ascospores; E, an ascospore very highly magnified.

up around these; its peridium is smooth and composed of a thin pseudoparenchymatous layer. In *Sartorya*, the cleistothecium is formed around an ascogonium, but no antheridia are present and the peridium is composed of several layers of loose, cottony hyphae. In *Emericella* there are no sex organs and the peridium is formed of several layers of interwoven hyphae with associated Hülle cells towards the outside; these curious cells have thick, strongly refractile walls, which in some species can be seen to be delicately radially striate if mounted in congo red stain (Fig. 57). They originate rather like chlamydospores, as terminal or intercalary swellings of peridial hyphae, and their thick walls are pierced by one or more pores where they are attached to the hyphae. However, the function of Hülle cells is still quite unknown. In *Hemicarpenteles*, the cleistothecium develops from a sclerotium which becomes unilocular within and produces ascogenous hyphae and asci at the centre; these eventually fill most of the cleistothecium, leaving the outer part of the sclerotial tissue as the peridium consisting of several layers of thick-walled cells. A somewhat similar form of development occurs in the species *Aspergillus alliaceus* (Fennell and Warcup, 1959; Fig. 11, B), where cleistothecia are formed as independent bodies within a sclerotium which requires several months before differentiation occurs.

The asci associated with perfect states of Aspergilli are broad, usually globose, ovoid or pyriform. They disintegrate at an early stage, leaving the ascospores free within the cleistothecia. The basic shape of the ascospores is like that of a pulley-wheel, but the two halves may be variously ornamented thus often providing a distinctive, specific character. There are eight or fewer ascospores per ascus, and they germinate by germtubes.

***Form-genus* Penicillium:** Members of the genus *Penicillium* (Fig. 58) are as ubiquitous and important as the Aspergilli. They are destructive in the same ways, but are useful in flavouring cheeses and in the preparation of some antibiotics, e.g. penicillin.

The conidiophore in *Penicillium* lacks a distinctive foot-cell, and it branches at the apex into a brush-like structure known as the **penicillus.** This is composed of conidia in basipetal chains, phialides, and sometimes one or more ranks of supporting branches. The penicillus is regarded as **monoverticillate** when the conidiophore ends in a whorl of phialides; and **biverticillate** if branching occurs at two (sometimes three or more) levels at the apex of the conidiophore before the level of phialides is reached. Biverticillate penicilli are termed **symmetrical** if the branches are evenly spaced about the central axis, and **asymmetrical** if they are lopsided. Primary branches from the conidiophore apex are known as **rami** (s. **ramus**); secondary ones as **metulae** (s. **metula**).

It will be observed that, basically, these structures are like those found in *Aspergillus*, except that in the latter the conidiophore apex is vesicular

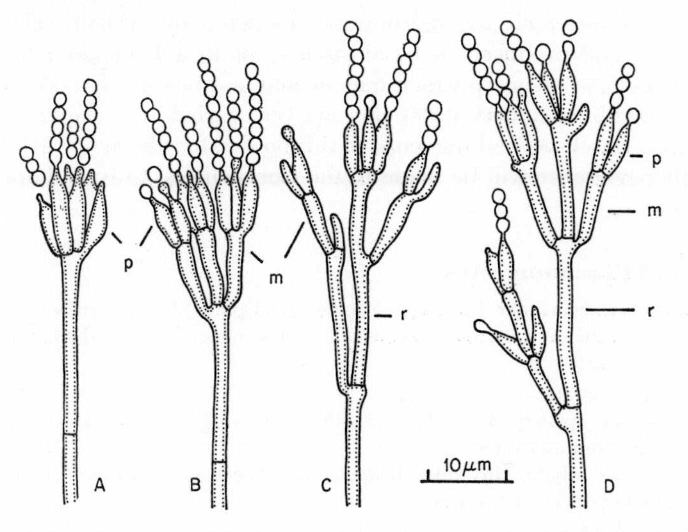

Fig. 58. *Penicillium* sp., conidiophores and conidia. A, monoverticillate penicillus; B, biverticillate symmetrical; C, D, biverticillate asymmetrical; *p*, phialide; *m*, metula; *r*, ramus.

instead of branching into rami. Intermediates are occasionally found, where a penicillus arises from a vesicle.

The sexual states of species of *Penicillium* have been placed in the genera *Eupenicillium* (Stolk and Scott, 1967) and *Talaromyces*. In *Eupenicillium* the cleistothecia originate as a sclerotium-like mass of thick-walled tissue which reaches a certain size and then starts forming ascogenous hyphae and asci from the centre outwards; the peridium is tough, and several cells in thickness. In *Talaromyces* (C. R. Benjamin, 1955) the cleistothecial wall is a loose weft of hyphae and the cleistothecium continues to grow even after the first ascospores have reached maturity. In general, the cleistothecia of Penicillia are firmer than those of Aspergilli, but are very similar. The ascospores are also pulley-wheel shaped.

CLASS PYRENOMYCETES

While the majority of Pyrenomycetes have true perithecia which are ostiolate and have their own peridial tissue even when embedded in a stroma, there are some orders in which the fruitbodies are cleistothecia or ascostromata. The Erysiphales have cleistothecia, but these forms are placed with the Pyrenomycetes rather than with the Plectomycetes because the asci arise at a common level towards the base of the cleistothecium; the Coryneliales and Coronophorales have ascostromata, a character of the Loculoascomycetes, but are placed with the Pyrenomycetes because their asci are unitunicate. Thus the Pyrenomycetes can be defined as Ascomycotina whose asci are unitunicate, inoperculate, usually clavate or cylindrical, and produced in basal fascicles or in a hymenium lining the inside of a closed ascocarp which may or may not be ostiolate. The conidial states formed by members of this class are very varied. An adequate account of the class is well beyond the scope of this book; after the key which follows, no further reference will be made to the Coryneliales, Coronophorales and Meliolales.

Orders of Pyrenomycetes

1. With cleistothecia: mycelium parasitic on aerial parts of higher plants, superficial and white or sometimes immersed; conidia usually abundant, white; asci indehiscent — *Erysiphales*
2. With ascostromata:
   A. Ascocarps top-shaped or spherical, often collapsing to become cup-shaped, with or without an ostiole — *Coronophorales*
   B. Ascocarps lobate, the lobes bearing asci becoming columnar and opening apically by a split or pore — *Coryneliales*
3. With perithecia:
   A. Parasitic on aerial parts of higher plants; mycelium superficial and dark, bearing hyphopodia — *Meliolales*
   B. Saprophytic or parasitic but with mostly immersed mycelium, lacking hyphopodia
      (*a*) Asci dissolving at maturity; ascospores liberated in a mucilaginous mass within the ascus — *Chaetomiales*
      (*b*) Asci persistent
         (i) Asci long, cylindrical, with scolecosporous thread-like ascospores — *Clavicipitales*
         (ii) Ascospores, if scolecosporous, then not thread-like
            (*m*) Perithecia or stromata light or bright coloured, soft, fleshy or waxy in texture — *Hypocreales*
            (*n*) Perithecia or stromata dark coloured, hard, membranous or carbonous in texture
               (*x*) Asci attached to the perithecial wall at maturity — *Sphaeriales*
               (*y*) Asci becoming loose in the perithecium at maturity, and liberated whole through the ostiole — *Diaporthales*

Order Erysiphales

These powdery mildews (Figs. 59, 60) are so-called because as obligate parasites on leaves, stems, calyces and fruits, their white mycelium and especially their masses of white conidia form a conspicuous powdery covering on the host. They form a neat, natural group of about a dozen genera. Various species attack a wide range of host plants. Some of the more notable are: *Erysiphe cichoracearum* on cucurbits and ornamentals; *E. graminis* on wheat and other cereals or grasses; *Sphaerotheca pannosa* on roses; *Podosphaera leucotricha* on apple; and *Uncinula necator* on grapevines.

In most genera, the mycelium forms an entirely superficial epiphytic network anchored to the host epidermis by appressoria and sending haustoria

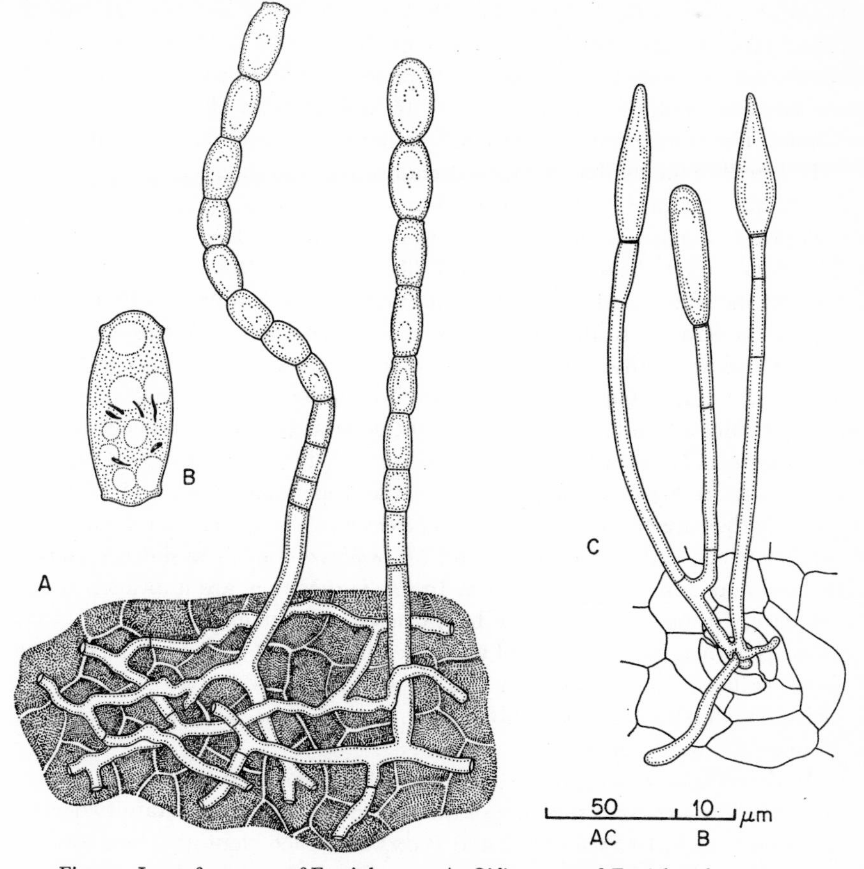

Fig. 59. Imperfect states of Erysiphaceae. A, *Oidium* state of *Erysiphe cichoracearum*; B, the same showing a conidium with prominent vacuoles and fibrosin bodies; C, *Oidiopsis* state of *Leveillula taurica*.

(Fig. 7) into the epidermal cells. The appressoria may be lobate or unlobed, and the haustoria vary from small, simple bulbs to large and digitately branched cells. The imperfect state of such genera is *Oidium*, a form-genus in which the conidiophore is unbranched and arises erectly from the surface mycelium, forming a basipetal chain of meristem aleuriospores at its apex (Fig. 59). The conidia are ellipsoid, but tend to be constricted near their ends when immature or when dried or mounted in certain mountants. They are often vacuolate and may contain highly refractile fibrosin bodies; they are formed in chains which may comprise as many as twenty-five or more conidia.

In the genus *Phyllactinia* the hyphae are mostly superficial, but some develop short branches which penetrate stomata and send haustoria into the mesophyll cells. *Phyllactinia* and *Uncinulopsis* have imperfect states belonging to the form-genus *Ovulariopsis*. The *Ovulariopsis* conidia are large, clavate to fusoid aleuriospores which arise singly at the apex of unbranched conidiophores which come from epiphytic hyphae. As the mature conidium falls off the conidiophore and is replaced by another, no chain of conidia is formed. The conidiophores are usually rather narrow and sparsely septate.

In *Leveillula* the hyphae are mostly endophytic and intercellular within the host tissues. Its imperfect state is the form-genus *Oidiopsis* (Fig. 59), in which endophytic hyphae emerge through the stomata and form branched conidiophores; these bear aleuriospores like those of *Ovulariopsis*, singly or forming short chains under humid conditions (Clerk and Ayesu-Offei, 1967).

In all genera, cleistothecia (Fig. 60) are formed on mycelia external to the host. These are whitish at first but soon change to yellowish, orange, brown and finally almost black. The peridium is composed of thickened polygonal cells forming a two-layered wall in most genera, but only one cell thick in *Brasiliomyces*. It usually bears appendages, which are a useful generic character. In *Erysiphe* and *Sphaerotheca* the appendages are hypha-like; in *Uncinula* they are circinately coiled; in *Sawadaea* their apices are recurved and bifid or trifid; and in *Podosphaera* and *Microsphaera* they form short, regularly dichotomous branches at the apex. In *Phyllactinia* they are unbranched and subulate, arising from a bulbous base whose wall is thicker in the distal than in the proximal part. Because of this difference in wall thickness, the bulbous bases collapse on drying and the appendages tend to lever the whole cleistothecium away from its substratum, thus perhaps aiding spore-dispersal. *Brasiliomyces* lacks appendages.

In the various genera, the asci are broadly clavate to subglobose, arise at a common level towards the base of the cleistothecium and contain two–eight ascospores each. In *Sphaerotheca* and *Podosphaera* each cleistothecium contains only one ascus; in the other genera there are several asci per cleistothecium.

The conidia are usually reckoned to be the summer dispersal spores and the cleistothecium a hibernating state of the fungus. Although this holds good in cold climates, it is usual in warm climates for the species to exist all the year round in the mycelial state and to produce conidia whenever there is a period of sufficiently wet or humid weather. In these circumstances cleistothecia may rarely, if ever, be formed by some species. Mycologists in hot climates are thus faced with the problem of classifying species of Erysiphales occurring mostly in their conidial states, whereas the most useful classifications are based on their cleistothecial states (Salmon, 1900; Blumer, 1933). Clare (1964), in Queensland, paid close attention to such indications as conidial-types, haustorial-types, appressoria, the presence of fibrosin bodies in conidia, and germtube-types and was able to use these features to identify many species from their asexual states alone.

Many species of Erysiphales are attacked by a fungal hyperparasite, *Ampelomyces quisqualis*, which forms small dark pycnidia, with ellipsoid

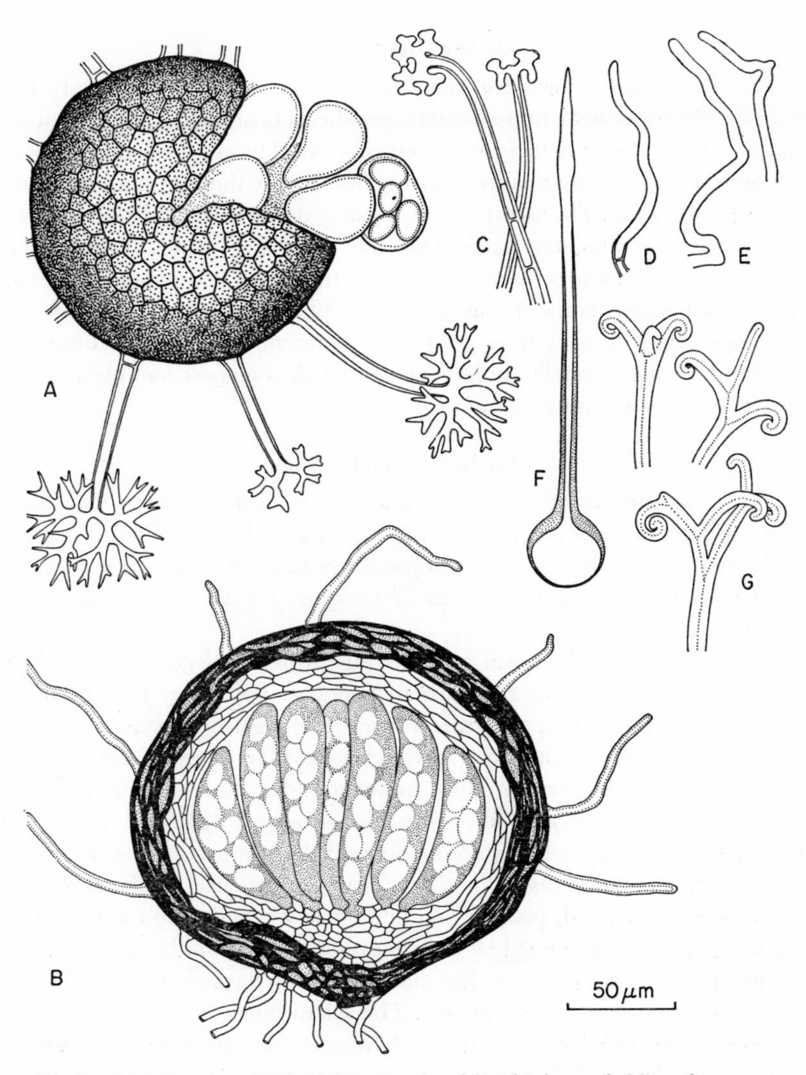

Fig. 60. Perfect states of Erysiphaceae. A, cleistothecium of *Microsphaera* sp. showing several asci emerging from the split pseudoparenchymatous peridium, and apically dichotomous appendages; B, section of cleistothecium of *Erysiphe aggregata*, showing asci in a basal fascicle and mycelioid peridial appendages. Appendages: C, *Podosphaera leucotricha*; D, *Sphaerotheca pannosa*; E, *Erysiphe aggregata*; F, *Phyllactinia corylea*; G, *Uncinula* sp.

hyaline unicellular conidia, within the conidiophores or conidia of the erysiphaceous host.

Order Chaetomiales

This is a small order with one family, Chaetomiaceae, and only three genera. The superficial, non-stromatic perithecia bear long, straight, coiled or branched brown hairs which are most distinctive. The asci dissolve away before the ascospores are fully mature, leaving them free within the glutinous matrix in the perithecium. The ascospores are usually dark brown and amerosporous.

The Chaetomiales, and particularly members of the genus *Chaetomium*, are important as saprobic cellulolytic fungi, rotting straw and paper, textiles and herbivorous dung. Some species are thermophilic and are able to grow well at temperatures approaching 50° C; their presence on straw in commercial mushroom beds is usually an indication that the compost had been allowed to overheat during its fermentation.

Order Clavicipitales

Two genera in the family Clavicipitaceae are of interest: *Claviceps* (Fig. 61) and *Cordyceps* (Fig. 62). Species of *Claviceps* are pathogens of various grasses and cereals, while those of *Cordyceps* parasitize caterpillars or insect larvae, or less commonly other fungi such as *Elaphomyces*, a hypogean member of the Plectomycetes.

In *Claviceps*, the ascospores are fine thread-like scolecospores which are dispersed by wind in spring just when grasses are flowering. They germinate and the germtubes infect the grass ovaries. The resulting mycelium replaces the ovary with a mycelial mat in which acervular layers of short conidiophores are formed. These produce innumerable small conidia and together constitute the asexual state known as the form-genus *Sphacelia*. At this stage the infection may not be very obvious except that the grass inflorescences are sticky to touch. The reason for this is that a sweet, sticky substance called 'honeydew' is secreted, possibly by the host as a reaction to the presence of the fungus. Insects attracted by the honeydew, carry conidia away and infect further host plants. The mycelial mat eventually hardens into a sclerotium whose common name is an '**ergot**'. The sclerotium is usually rather larger than the grain it replaces; it is hard and horny, often curved and with a dark purplish-black rind; it falls to the ground and overwinters before germinating. On germination, each sclerotium may bear one or more stipitate stromata. These are subspherical and contain perithecia immersed just below the surface. Each perithecium has a peridium consisting of several layers of cells, is ostiolate, and has periphyses lining the interior of the ostioles. The asci are long and narrow, cylindrical with a thickened truncate apex, and each contains eight thread-like ascospores arranged parallel to one another.

The sclerotia contain several alkaloids which may poison stock feeding on

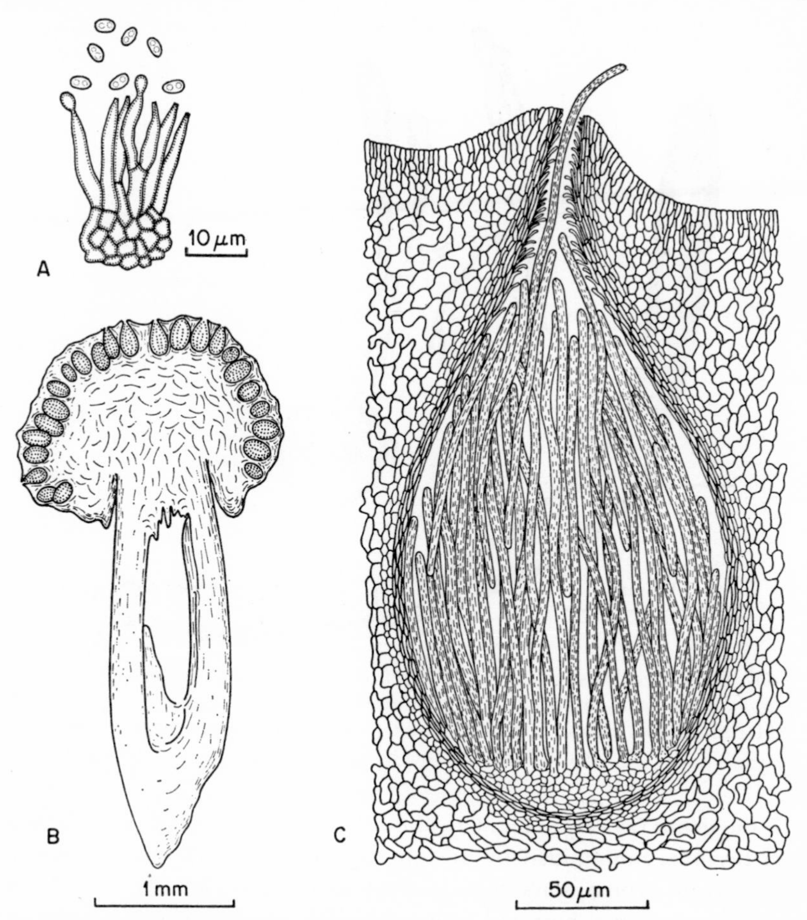

Fig. 61. *Claviceps purpurea*. A, conidiophores and conidia of the *Sphacelia* imperfect state; B, stipitate-capitate stroma (from a germinated sclerotium) showing distribution of perithecia in periphery of the stroma; C, a perithecium embedded in stromatic tissue, showing the peridium, periphyses lining the ostiole, asci including one emerging from the ostiole.

infected cereals, with symptoms ranging from lack of muscular co-ordination to abortion. *Claviceps purpurea*, which occurs principally on rye, is particularly toxic, and humans are sometimes poisoned, with symptoms of gangrene, by eating rye bread in which sclerotia have been milled with the grain. Ergot alkaloids are extracted pharmaceutically and used in obstetrics and in the treatment of migraine, vascular diseases and high blood pressure. The production of alkaloids varies considerably with the species of *Claviceps* and also with climate; it has been found, for instance, that it is commercially unprofitable to extract alkaloids from ergots in warm climates.

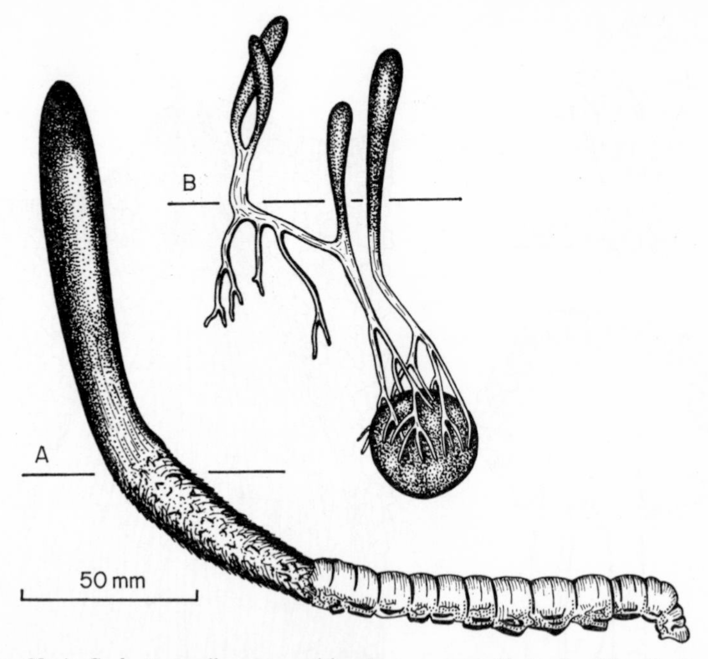

Fig. 62. A, *Cordyceps gunnii*, stroma arising from a mummified caterpillar; B, *Cordyceps* sp., stromata arising from a parasitized ascocarp of *Elaphomyces muricatus*. Horizontal lines indicate soil level. (A, after a painting by Gwen Walsh presented to the Waite Institute by Sir John Cleland.)

In the genus *Cordyceps* (Fig. 62), the body of the host caterpillar or insect larva becomes mummified by an internal mat of mycelium. Eventually a large, clavate stroma, sometimes as much as 10–15 cm in length, grows out from the mummified host and bears immersed perithecia in its upper part. The perithecia, asci and ascospores are very like those of *Claviceps*. *Cordyceps gunnii*, a species with very large stromata occurring on caterpillars in Australia and New Zealand, was at one time used by the Chinese for its reputed medicinal properties. *Cordyceps ophioglossoides* is interesting as a parasite of other fungi; it occurs on the hypogean ascocarps of species of *Elaphomyces*, forming its stromata above ground level.

Order Hypocreales

The more notable families of this order are the Nectriaceae and the Hypocreaceae.

Family Nectriaceae

The perithecia in the Nectriaceae may be discrete or are formed superficially in groups on a flat subicular stroma, and usually they are reddish or orange. The perithecium has an ostiole with periphyses, and contains pseudoparaphyses. The ascospores are hyaline and two-celled.

Of the several genera in this family, *Gibberella* is of interest because the series of plant hormones known as 'gibberellins' were first isolated from some of its species pathogenic on grasses and cereals. The genus *Nectria* contains many species which are serious pathogens of trees. *Nectria galligena* is well known as the cause of apple canker, but attacks other fruit and shade trees as well. *Nectria cinnabarina* is a less serious pathogen of shade trees and of some species of *Ribes*. The conidial states of various species of *Nectria* are distributed among fourteen form-genera in four form-families of the Deuteromycotina; all, however, are forms with phialides and phialospores (Booth, 1959). *Nectria galligena* forms white sporodochia bearing large curved phragmospores of the *Cylindrocarpon* type; the imperfect state of *N. cinnabarina* consists of pink erumpent sporodochia with small unicellular conidia, and is referred to the form-genus *Tubercularia* (Fig. 48). Certain other species of *Nectria* have imperfect states belonging to the form-genera *Fusarium* and *Verticillium*.

Family Hypocreaceae

In *Hypocrea* the stroma is thick and pulvinate (cushion-shaped). Perithecia are formed within the stroma, with only the short necks of the ostioles protruding from its surface. The ascospores are two-celled but become fragmented into unicellular part-spores. Conidial states of some of these species belong to the form-genus *Trichoderma* which is exceedingly common on decaying wood and as a laboratory weed, often making it difficult to isolate wood-destroying fungi of other types in pure culture.

Order Sphaeriales

The Pyrenomycetes were, for a long time, divided into only two orders: the Hypocreales with much the same circumscription as adopted in this text, and the Sphaeriales (Fig. 63) for the remainder. Thus it will be noted that the name Sphaeriales is currently applied to a much more restricted group than it was formerly. It is still a large order, in which mycologists recognize different numbers of families. We shall not be able to do more than touch briefly on a number of the more interesting genera and species in this account.

The perithecia of Sphaeriales may be discrete or formed in stromata, but always have a well defined peridium of their own and always have periphyses lining the ostioles. Most Sphaeriales are saprobic on such materials as wood, paper, straw or dung, but there are also a number of important pathogens. Some of the saprobic species have been used extensively in biochemical and genetical studies; these include *Sordaria fimicola* and especially *Neurospora crassa* and *Neurospora sitophila*, the last being the pink bread-mould which sometimes causes trouble in bakeries and can be a vicious contaminant in laboratories. Some of the other saprobes cause blue-stain of timber, or soft-rot of water-soaked timber used as slats in cooling towers of electric power stations. The pathogenic members of the Sphaeriales may attack living trees (e.g. *Rosellinia*, *Nummularia*, *Eutypa*, *Valsa* and *Diatrype*) sometimes causing cankers, or they may parasitize herbaceous plants such as grasses (e.g. *Phyllachora*).

The stromata in stromatic Sphaeriales are sometimes sufficiently distinctive to be given a common name, such as 'valsoid', 'eutypoid' or 'diatrypoid', corresponding with the types found in the genera *Valsa*, *Eutypa* and *Diatrype*, respectively (Fig. 63); but more often they are variable within a particular family. In the Xylariaceae, for example, the stromata characteristic of different genera may be crust-like, dome-shaped, club-shaped or coralloid.

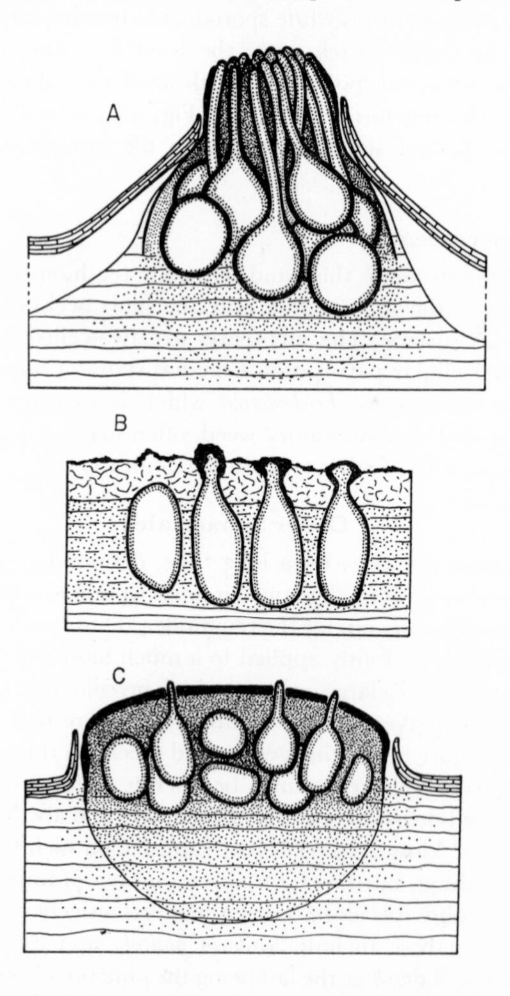

Fig. 63. A valsoid; B, eutypoid; and C, diatrypoid perithecial stromata.

Order Diaporthales

Several genera of this order have species which are notable as plant pathogens: e.g. *Gnomonia* in the Gnomoniaceae and *Diaporthe*, *Endothia* and *Glomerella* in the Diaporthaceae. Especially noteworthy is *Endothia parasitica*, the cause of chestnut blight, which virtually wiped out the native chestnut in North America when it was introduced from Asia early this century. *Glomerella cingulata* and its imperfect state, *Gloeosporium fructigenum*, are the cause of bitter-pit disease of apple, while *Glomerella lindemuthiana*, with a *Colletotrichum* imperfect state, causes anthracnose disease of French beans. The term 'anthracnose' refers to a disease characterized by shallow, limited lesions which bear acervuli of a member of the Melanconiales (Deuteromycotina).

CLASS DISCOMYCETES

The Discomycetes include most of the larger types of Ascomycotina such as cup-fungi, earth-tongues (Geoglossaceae), morels (*Morchella*) and truffles (Tuberales). The fruitbodies are simple or compound apothecia varying greatly in size, colour, shape and construction. They are not necessarily discoid or cupulate but may instead be bell-shaped, saddle-shaped, spathulate, or cerebriform (convoluted like a brain). Most of these more unusual shapes are associated with stipitate and pileate fruitbodies, while the cupulate ones may be sessile or stipitate. All these types of ascocarps bear the asci in hymenia in open cavities or at an open plane surface, except in the case of the truffles. The ascocarps of truffles are formed underground and in this situation remain more or less enclosed until disintegrated by weathering or by small animals; in extreme forms, the asci tend to be scattered and embedded in more or less solid ascocarp tissue. A series of connecting forms exists, which shows that the truffle fruitbody is almost certainly derived from apothecial Discomycetes and probably from forms with operculate asci. In the hypogean environment, however, the asci no longer have a liberation and dispersal function and have become indehiscent.

Orders of Discomycetes

1. Asci operculate; ascocarps epigean (Operculate Discomycetes) — *Pezizales*
2. Asci inoperculate, dehiscing by an apical pore or slit (Epigean Inoperculate Discomycetes)
   - *A.* Asci long, narrowly cylindrical, with thread-like ascospores — *Ostropales*
   - *B.* Asci not narrowly cylindrical; ascospores not thread-like even when scolecosporous
     - (i) Apothecia immersed in substratum and covered at first with a radiately fissured stroma — *Phacidiales*
     - (ii) Apothecia superficial, not covered with a stroma — *Helotiales*
3. Asci indehiscent; ascocarps hypogean, enclosing or almost enclosing more or less solid tissues (Hypogean Discomycetes) — *Tuberales*

Order Pezizales

Dennis (1960) recognizes six families of the Pezizales or operculate Discomycetes. The ascocarps in the Morchellaceae and Helvellaceae have, in most instances, a well defined stipe expanded towards the apex into a distinct pileus which may be smooth or pitted and ridged. In the four other families, Pezizaceae, Humariaceae, Sarcoscyphaceae and Ascobolaceae, the apothecium is typically discoid or cupular, usually sessile, but sometimes shortly stipitate.

In *Helvella* ('false morel'; Fig. 64) there is a somewhat saddle-shaped pileus with a smooth or convoluted hymenium on the upper side; one species, *H. esculenta*, might, by its name, be supposed to be edible but contains

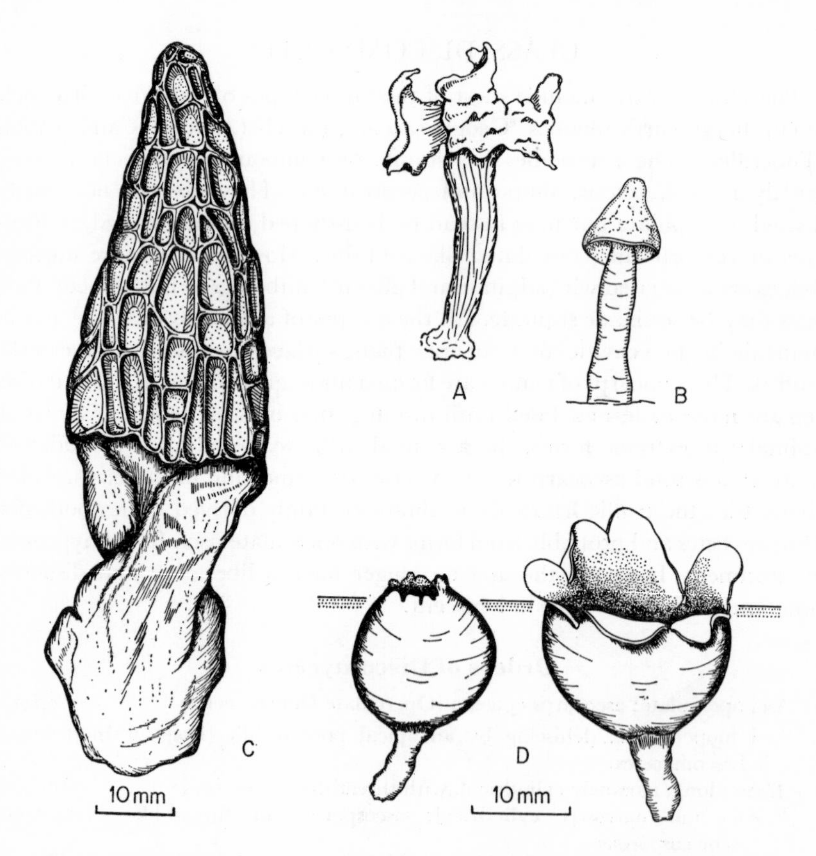

Fig. 64. Some Pezizales. A, *Helvella crispa*; B, *Verpa* sp.; C, *Morchella* sp.; D, *Sarcosphaera ammophila*.

helvellic acid which is poisonous, at least to some people. *Rhizina* is placed in the Helvellaceae although its apothecia are undulating and discoid and attached to the soil by several tough root-like structures; *R. undulata* is found in cleared or burnt areas of pine forests and has been implicated as a pathogen of pine seedlings.

Species of *Morchella* ('morel'; Fig. 64) have a pitted or ridged pileus borne on a stout stipe; the hymenium lines the cavities between the network of ridges. In section it can be seen that each cavity represents an apothecium and that these are compounded at the surface of the rather large, hollow pileus. Species of *Morchella* are edible and much valued for their flavour. Unfortunately, no-one has yet succeeded in cultivating them, although cultures of mycelium are not difficult to establish on agar. Another genus of the Morchellaceae is *Verpa* (Fig. 64), whose members have a bell-shaped, conical or cylindrical pileus which surrounds the stipe, but is free from it, and has a smooth hymenium on the upper surface.

Most of the Pezizaceae tend to have rather large apothecia, usually more than about 1 cm in diameter and sometimes as much as 10–15 cm. The central genus of this family is *Peziza*, with typical apothecia occurring above ground on such substrata as soil, dung, plaster and plant debris. The ascocarps in the genus *Sarcosphaera* (Fig. 64) are formed in the surface layer of soil and at first are spherical and immersed; later the apex of the apothecium bursts through the soil and ruptures stellately to expose the hymenium in the deeply cupular, large fruitbody. In sandy soil, the apothecia look not unlike holes made by a walking-stick.

Most of the smaller types of apothecia belong to members of the remaining families. The apothecia in the Humariaceae are small, often brightly coloured and usually beset with hairs on the outside. Members of the Ascobolaceae inhabit dung, burnt soil or plant debris; their mature asci usually protrude above the level of the immature asci in the hymenium, and their spores are usually biseriate or scattered within the ascus. *Pyronema omphalodes*, which has already been mentioned in connexion with the development of a typical apothecium, is a member of the Ascobolaceae. Most of the Pezizales are saprobes.

Order Tuberales

These hypogean Discomycetes occurring often in association with roots of higher plants, are known by the common name 'truffles' (Fig. 65). The ascocarps are formed underground, often several centimetres below the surface of the soil. Some have retained a more or less cupular form but do not open widely, while others have a hollow interior communicating by small fissures to the outside; still others have a completely closed ascocarp whose interior is a solid fleshy tissue with a marbled or veined pattern caused by irregular convolution of sterile and fertile parts. The asci, which are embedded

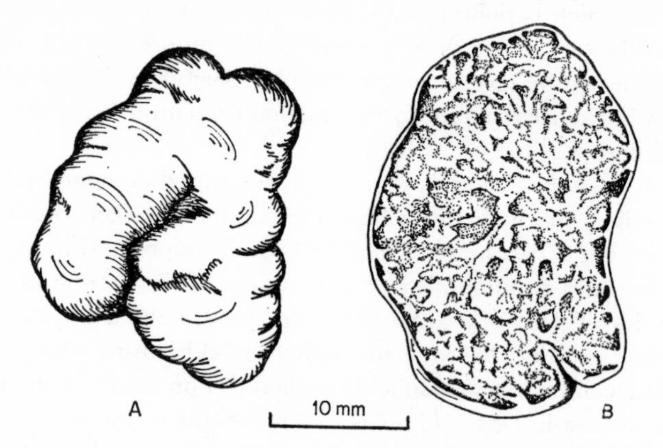

Fig. 65. *Tuber clarei.* A, external view of an ascocarp; B, internal view of the same showing irregularly veined areas.

in this fleshy stroma, are usually spherical or broad and indehiscent, while the ascospores tend to be spherical or broadly ovate, often thickened and ornamented with warts or reticulate ridges. The exterior of the ascocarp may be smooth and whitish in some species, or thick, black and warted in others. Members of the Tuberales can easily be confused at first sight with some of the hypogean Gasteromycetes, but are readily differentiated when examined microscopically.

The truffles are edible and are considered a delicacy in most countries where they occur. In nature they are probably dispersed by rodents. In Europe, pigs and dogs are specially trained to hunt the underground fruit-bodies of *Tuber* species by their scent. In the Kalahari desert of South Africa, where about three species of *Terfezia* occur, the Bushmen have become adept in detecting the presence of these fungi by observing minute cracks in the soil-surface above the expanding fruitbodies, which may be deeply buried. In Central Australia a species of *Elderia* occurs near Alice Springs and is eaten by the Aborigines. Although not all Tuberales occur in desert regions, it is notable that many do; they appear only after rain has fallen.

The question will certainly be asked: why are the Tuberales considered to be Discomycetes, when these have an open hymenium at maturity while truffles do not? This may best be explained by considering a series of generic types which show increasing convolution of the hymenium accompanied by closing of the apothecium until finally a solid, closed fruitbody is evolved. I do not suggest that one generic type has been derived directly from another, but merely that they illustrate the type of change that must have occurred in the evolution of the truffles. We have already noted that some Disco-mycetes, such as *Sarcosphaera* (Fig. 64) are semi-hypogean, but that their apothecia open through the surface of the soil before they mature. Some of the Tuberales (e.g. *Hydnocystis*) are not unlike *Sarcosphaera*, but are truly hypogean, with a cupular or hollow ascocarp whose smooth hymenium contains asci and paraphyses. Various further lines of development have been sketched by Gilkey (1939). In certain genera, the tips of paraphyses projecting beyond the asci have apparently fused to form a secondary tissue which encloses the asci in intercommunicating chambers, some of which may open independently at the surface of the ascocarp. In addition, strands of sterile tissue may grow up through the hymenium and connect with the secondary tissue beyond the asci, thus increasing the number of isolated chambers formed. In other genera, however, the course of development has apparently been different. The hymenium has become increasingly convoluted with folds, pleats and ridges which eventually more or less fill the cavity of the ascocarp and coalesce, so that the interior becomes divided into chambers or tortuous interleading channels lined by a hymenium. In some such genera there may be one or more small openings leading to the surface of the ascocarp, but finally, completely closed ascocarps are produced. If we suppose that convolution of the hymenium has evolved because pressure of the surrounding soil limits lateral expansion of the ascocarp in a hypogean situation, then a closed, virtually solid ascocarp is to be expected as the end-point. At the same time, convolution is a means of greatly increasing the spore-bearing area of the hymenium without unduly increasing the bulk of the fruitbody.

The Tuberales, then, probably represent a special adaptation of Operculate Discomycetes to the hypogean habitat; their large ornamented amerosporous ascospores suggest their derivation from the operculate group, even though the asci have become inoperculate (indehiscent) as a result of loss of their normal function of liberating the ascospores.

Orders Ostropales, Phacidiales and Heliotales

It is not possible to do justice to the vast number of epigean inoperculate Discomycetes in an account of this sort; the reader is referred to Dennis (1949, 1956, 1960) for a comprehensive and modern treatment. Three orders are usually recognized: Ostropales, Phacidiales and Helotiales, the last being possibly the largest order of all Ascomycotina.

In the Ostropales, the ascospores are thread-like and eventually break down into part-spores; the ascal walls are thickened and canaliculate at the apex. These characteristics are similar to those found in the Clavicipitales (Pyrenomycetes). The Phacidiales are recognized by their peculiar ascocarps which are developed within host plant tissue and eventually erupt through it. The hymenium is covered by layers of black sclerenchymatous stroma which becomes radially or longitudinally fissured and at length weathers away to expose the hymenium. In *Rhytisma* (Fig. 66) several hymenia are present within and beneath the common radiately fissured stroma; species of *Rhytisma* cause 'tar-spot' disease of leaves of maple and willows. In

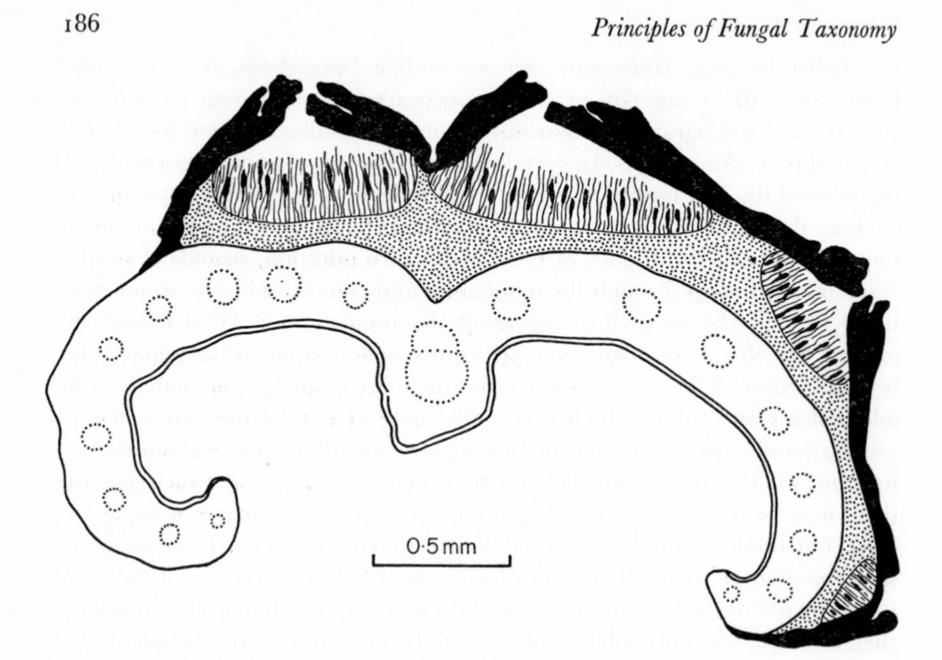

Fig. 66. *Rhytisma andromedae*, section of ascocarp showing four hymenia in a common stroma covered by a fissured black stromatic layer (host leaf in outline only).

Lophodermium and *Hypoderma*, which are agents of needle-cast of conifers, the ascocarps are not unlike hysterothecia in shape and each contains a single hymenium. The ascocarps of Phacidiales are not typical apothecia but they qualify in this category because the hymenia are eventually exposed.

Order Helotiales

The asci of members of this order are only slightly thickened at the apex and the ascospores are usually not scolecosporous. The ascocarps are discoid to cupular and mostly small. While most of the Helotiales are saprobes, a few are serious pathogens of plants. Dennis recognizes seven families in this order, of which we shall consider three.

Family Sclerotiniaceae

From an economic point of view this is a most important family, whose members are responsible for much damage to fruit, vegetables and ornamentals. The apothecia are cupulate, usually borne on relatively long stipes which arise from sclerotia (Fig. 11) or stromatic tissue after spermatization of receptive cells on these structures. The ascospores are hyaline, unicellular, oval or somewhat elongated. Whetzel (1945) has divided the family into fourteen genera, some of which have characteristic conidial states; one of the latter is *Botrytis*, some of whose species are imperfect states of species of *Botryotinia*. *Botrytis cinerea* is extremely common as 'grey mould' on a very wide range of host plants, where it probably occurs as a secondary parasite in most instances; this species, probably an aggregate species, was the subject of Brown's (1915, 1916, 1917) classical investigations on the physiology of parasitism.

Monilinia fructicola is the cause of peach mummy disease and brown-rot of other stone fruits. Flowers of susceptible hosts are infected in spring by mycelia originating from the germination of conidia or ascospores. The conidial state (form-genus *Monilia*) has long conidiophores whose branches terminate in chains of ellipsoid blastospores (Fig. 44, B) which are the means of disseminating the fungus during summer. The peach fruit first shows a brownish lesion on the outside and becomes rooted inside by pectolytic enzymes which break down the pectates of the middle lamellae of the cells. Eventually, mycelium replaces most of the tissues of the fruit, which becomes mummified, dry and shrivelled as a pseudosclerotium. The surface of the fruit may then become covered with conidiophores and conidia. The mycelium overwinters in the dried fruit which usually becomes partly buried in soil. After 1–3 years the apothecia are formed from the pseudosclerotium. Spermatia are found and are probably responsible for the initiation of apothecia, although this has not been proved.

Family Geoglossaceae

Members of this family are saprobic in soil and are characterized by the production of stipitate ascocarps which expand toward the apex into a clavate or spathulate or flabellate fertile part bearing the hymenium on its outer surface. Fruitbodies of *Geoglossum* are clavate or spathulate, with hyaline or brown cylindrical ascospores usually divided into several cells. *Trichoglossum* (Fig. 67) is similar but has sterile brown **setae** projecting from the hymenium, as well as recurved paraphyses which tend to form an epithecium above the level of the asci. *Leotia* (Fig. 67) has a stipitate ascocarp with a gelatinous, lobed pileus on whose outer surface the hymenium is borne.

Family Cyttariaceae

Cyttaria (Fig. 68), a genus with few species, occurs on *Nothofagus* trees in New Zealand, Tasmania, Queensland, Chile, and Tierra del Fuego. Infected areas of host twigs become hypertrophied and gall-like, and from these clusters of ascocarps are eventually produced. Each ascocarp is whitish, more or less spherical but narrowing towards its point of attachment at the base. It is hollow and may be about as big as a golf ball. The entire outer surface is indented by deep polygonal cavities, each of which is lined by hymenial tissue and represents an apothecium. The contiguous apothecia are joined by narrow bands of sterile tissue. Thus the entire ascocarp may be viewed as

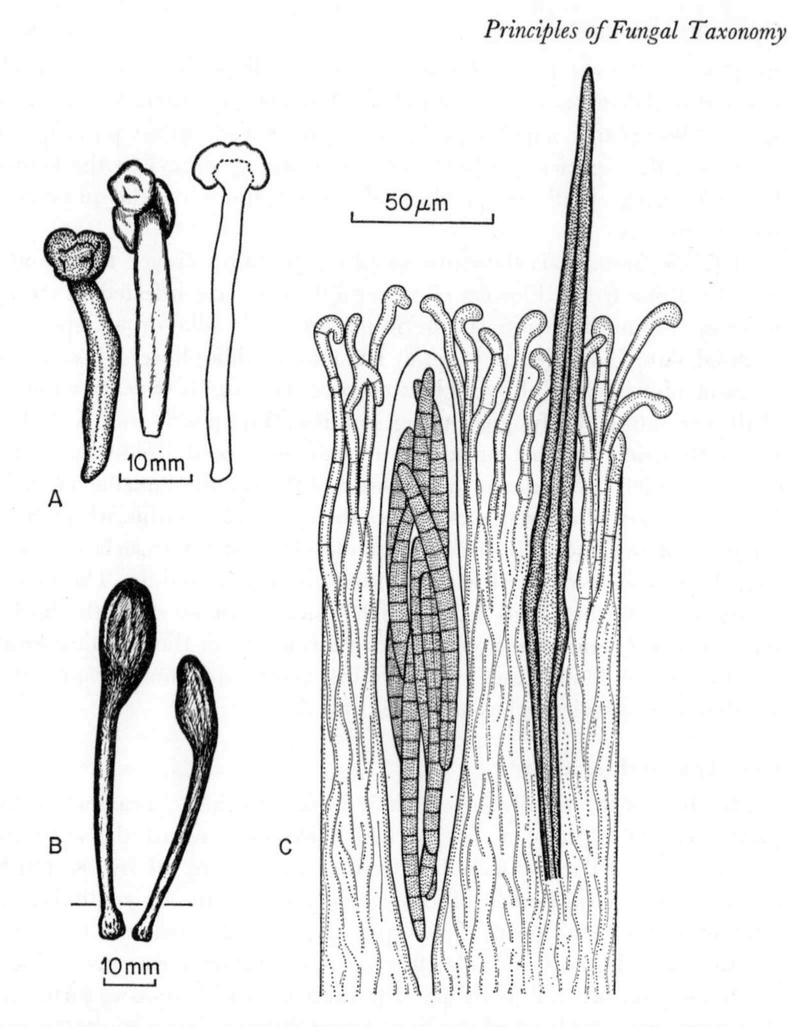

Fig. 67. Geoglossaceae. A, *Leotia lubrica*, stipitate-capitate ascocarps; B, *Trichoglossum hirsutum*, ascocarps; C, *T. hirsutum*, section through hymenium showing one unitunicate inoperculate ascus with brown ascospores, one seta, and recurved paraphyses.

a compound apothecium whose parts are connected by, and partly embedded in, fleshy stromatic tissue. The ascocarps are edible.

CLASS LOCULOASCOMYCETES

Fungi in this class are distinguished by having bitunicate asci borne in ascolocules in an ascostroma. In some instances, however, the true nature of the ascocarp cannot be determined without detailed developmental study.

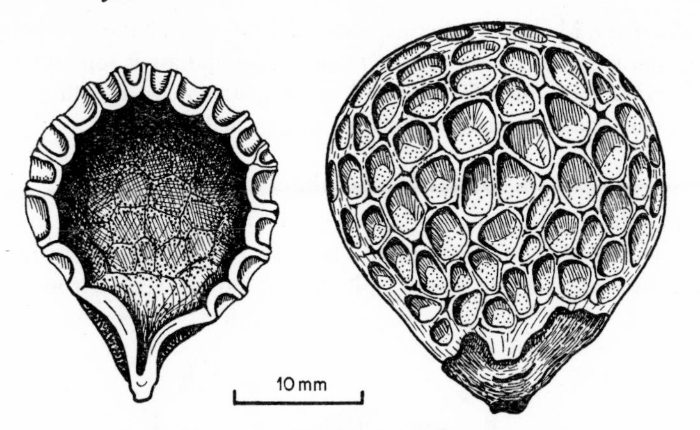

Fig. 68. *Cyttaria gunnii*, ascocarp in section (left) and external view (right).

For instance, some Loculoascomycetes have unilocular ascostromata (also known as **pseudothecia**) which may be difficult to distinguish from true perithecia, while some of the Discomycetes have apothecia which contain bitunicate asci. Certain other fungi have unitunicate asci in what appear, without close study, to be ascostromata. It should therefore be stressed that a combination of bitunicate asci with ascostromata is necessary to define a Loculoascomycete. Various well defined types of centrum development have been recognized in this class (Luttrell, 1951, 1955, 1965).

Orders of Loculoascomycetes

1. With uniascal locules formed at various levels in a cushion-shaped ascostroma or in a gelatinous thallus — *Myriangiales*
   With multiascal locules formed at a common level near the base of a plurilocular ascostroma, or with a unilocular ascostroma containing many asci — 2.
2. With a flattened, radiating shield-like ascostroma (thyriothecium) formed on a superficial mycelium or rarely subcuticular — *Hemisphaeriales*
   Ascostroma not flattened and radially constructed — 3.
3. Ascostroma a hysterothecium; asci arising among pseudoparaphyses — *Hysteriales*
   Ascostroma not a hysterothecium, or if resembling one then not containing pseudoparaphyses — 4.
4. Ascostromata small spherical uniloculate pseudothecia formed on a superficial mycelium which is often dark and conspicuous — *Capnodiales*
   Ascostromata not on a dark superficial mycelium — 5.
5. Pseudoparaphyses absent — *Dothideales*
   Pseudoparaphyses present — *Pleosporales*

Order Myriangiales

Various species of *Elsinoë* are plant pathogens, causing such diseases as citrus scab and anthracnose of grape, raspberry and avocado. All species of *Elsinoë* appear to have the same type of conidial state referable to the form-genus *Sphaceloma*. Many pathogenic species of *Myriangium* have been recorded; some are parasites of insects. In the tropics, human hair is sometimes parasitized by *Piedraia*, which forms small black gritty ascostromata round the hairs; the cure usually consists simply in cutting the hair off.

Order Hemisphaeriales

Most members of the Hemisphaeriales (Fig. 69) occur as leaf parasites or as epiphytic saprobes in the tropics. They are recognized primarily by the radiating structure of the outer wall of the flattened ascostroma, which

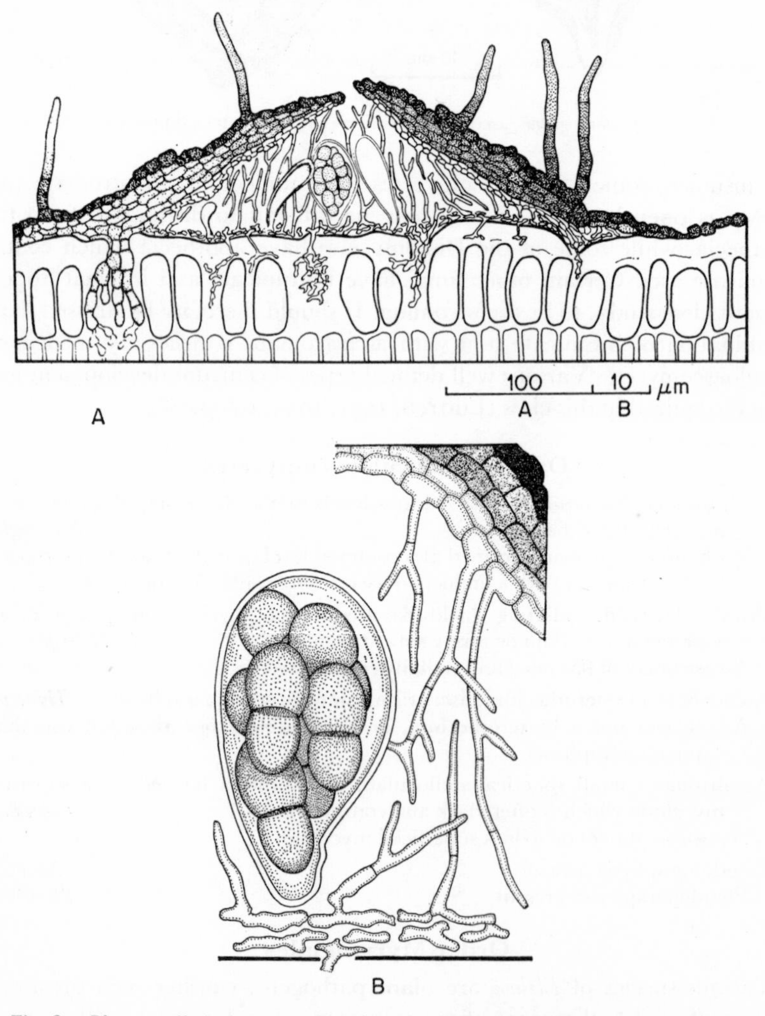

Fig. 69. *Placoasterella baileyi*. A, uniloculate thyriothecium (ascostroma) in section; B, a bitunicate ascus and pseudoparaphyses.

usually dehisces by means of stellate fissures splitting from the centre outwards. These ascostromata may be uniloculate or pluriloculate. Some forms, sometimes placed in a separate order, appear to have inverted ascocarps, with the asci developing from the inner tissues of the radiate covering of the ascostroma.

Order Hysteriales

The characteristic hysterothecia (Fig. 14, D, E) of members of this order are black and usually brittle, carbonous and elongated, with the upper surface bisected longitudinally by a slit-like opening in the overlying stromatic tissue. The asci are clavate to cylindrical and are accompanied by pseudoparaphyses. These ascostromata occur commonly on newly dead wood and may often be found on the spines of some of the thorny acacias in Africa.

Order Capnodiales

Commonly known as 'sooty moulds', the members of this order are mostly epiphytic saprobes associated with the secretions of insects, but some are hyperparasites or leaf parasites. Even the saprobic species may interfere with photosynthesis by covering a great area of the leaves of the host with their conspicuous shaggy mats of dark superficial mycelium. In some cases the host is penetrated by haustoria and an internal mycelium. The ascostromata are small, unilocular and more or less spherical; they open by a pore or by wearing away of the outer tissues. Sooty moulds appear to be particularly common in the Amazon jungles and in areas bordering the Pacific Ocean.

Order Dothideales

The ascostromata of Dothideales vary from uniloculate to pluriloculate, and cushion-shaped to perithecioid; the locules are small, spherical and ostiolate. The asci are ovate to clavate or cylindrical and are not associated with paraphyses or pseudoparaphyses.

Various species of *Mycosphaerella* cause leaf-spot diseases of pea, strawberry, pear and banana plants. A species of *Guignardia* is an important pathogen of grapes, while another causes black-spot of citrus. Spermatia borne in spermogonia are present in several species in the Dothideaceae, and are probably concerned in plasmogamy.

Order Pleosporales

The ascostromata in this order range in form from cushion-shaped plurilocular to unilocular perithecioid forms. The asci are clavate or cylindrical and occur, with pseudoparaphyses, in rather large spherical ostiolate locules. Both saprobes and parasites are included in members of this order.

The order is divided into several families, of which the Venturiaceae, Pleosporaceae and Lophiostomataceae are perhaps the best known. *Venturia inaequalis* and its conidial state, *Spilocaea pomi*, cause apple scab disease and also scab of some related fruit trees. Several genera in the Pleosporaceae contain species which are notable plant pathogens, e.g. *Leptosphaeria*, *Pleospora*, *Cochliobolus*, *Cucurbitaria* and *Pyrenophora*.

LICHENS

As no other mention will be made of lichens in this book it is perhaps appropriate to discuss them briefly at this point.

A lichen is a composite organism whose thallus consists of a fungal and an algal component in intimate association, while its fruiting structures are always those of the fungus. The fungus is most often one of the apothecium-forming Ascomycotina, less often perithecium-forming, and very rarely one of the Basidiomycotina. Any one 'species' of lichen always has the same fungal and algal components, forming a thallus of distinctive shape and quite unlike the inconspicuous mycelial thallus of a non-lichenized higher fungus. Because the thallus is distinctive it is described as if it belonged to a single organism. Thus it has been both traditional and convenient to study lichenology separately from mycology and algology. Some attempts have been made to incorporate the lichenized fungi into mycological classification, but in view of the vast number of species and genera concerned there are great problems to be overcome. Perhaps the most important of these is that changes in nomenclature would be unavoidable, since the existing names of lichens apply to the composite organisms and not to their fungal components alone.

The exact physiological relationship between the partners in a lichen association is not well known. The algal components can exist independently in nature, but the fungi apparently cannot do so readily although they can sometimes be grown in culture. Experiments to reconstitute the lichen from cultures of its components have given useful information on their delicately balanced relationship. Such work seems to show that the fungus is parasitic on some of the living algal cells, protects and possibly nourishes others, and lives saprobically on still other algal cells which it has killed or which have died from other causes.

Most lichens are xerophytes and are to be found especially as colonizers of bare rock and on the bark of trees. Certain dyes and indicators are still extracted from species of *Roccella*, while in the Arctic regions species of *Cetraria* and *Cladonia* are important sources of food for man and reindeer, respectively.

12

FUNGI WITH BASIDIA AND BASIDIOSPORES USUALLY IN FRUITBODIES: EUMYCOTA SUBDIVISION BASIDIOMYCOTINA

Although some Basidiomycotina exist mainly in the form of mycelia or as small fructifications on wood or litter, or under clods of soil, the majority form conspicuous fruitbodies (**basidiocarps**). Since these are often noticed and collected they have been given common names which describe their morphological forms. For instance, we speak of the rusts and smuts, jelly fungi, club fungi, brackets, mushrooms, puffballs, phalloids and earthstars. But the various fungi having any one of these forms of fruitbody are not necessarily related, and indeed some of these forms are encountered among Ascomycotina as well. Such fruitbodies are all examples of the phenomenon of **homoplasy,** the production of similar morphological forms along several unrelated phyletic lines. The different types of fruitbody, then, represent various solutions to common problems of existence in terrestrial fungi, especially those relating to efficient liberation and dispersal of spores. In passing it may be noted that only one or two species of Basidiomycotina are known to occur in water, and they do not form basidiocarps.

Apart from the presence of basidia, of which there are many morphological types, there are few features in common among all the different sorts of Basidiomycotina. In most instances the mycelium is extensive, though usually inconspicuous, and hyphae encrusted with crystals of calcium oxalate are not uncommon. Hyphal strands or rhizomorphs are often found. The septal characteristics of the hyphae have already been reviewed (p. 51), as have the cytological events leading to the production of basidiospores

(p. 97). Clamp connexions are present in the dicaryotic hyphae of a large proportion of species; these will be described shortly.

MYCELIUM, HYPHAE AND HYPHAL SYSTEMS

Although variations in pattern are encountered, generally the mycelium passes through three phases (primary, secondary and tertiary) during the development of the fungus.

A **primary mycelium** results from germination of the basidiospores and is composed of haploid monocaryotic cells. A **secondary mycelium** arises by dicaryotization of cells of the primary mycelium, by spermatization or somatogamy. In the Uredinales, dicaryotization takes place by fusion of spermatia with compatible receptive hyphae; in the Ustilaginales, sporidia (or basidiospores) may fuse and form dicaryotic cells; in most other instances anastomosis of somatic hyphae is responsible for dicaryotization and an extensive secondary mycelium is usually formed. Sometimes, as in *Thanatephorus cucumeris* (Flentje, Stretton and Hawn, 1963) the primary mycelium is haploid and multinucleate, and during cell division the number of nuclei per cell is progressively reduced until in the prebasidial cells there are only two nuclei.

A **tertiary mycelium** is also dicaryotic but composes the organized and specialized tissues of the basidiocarp. In some of the higher Basidiomycotina, particularly the Aphyllophorales, the hyphae of the tertiary mycelium may be differentiated into as many as three different **hyphal types** combined in three different **hyphal systems** (Corner, 1932*a*, 1932*b*).

Hyphal types and hyphal systems

The basic hyphal type is a **generative hypha** (Fig. 70). This is usually (but not always) thin-walled, much branched and septate, with clamp connexions in those species which have clamps. Any other hyphal types, basidia, and most sterile hyphal elements associated with the basidia, arise from generative hyphae. In some genera of clavarioid and agaricoid fungi with the more fleshy types of fruitbody, the generative hyphae are able to become turgid and inflated, thus assisting in supporting the erect fructification.

Skeletal hyphae are unbranched or very sparingly branched, usually relatively wide, with thickened walls and often an almost occluded lumen. They are often coloured, always lack clamp connexions, and if any septa are formed they appear to be adventitious.

Binding hyphae are branched at short intervals to form intricate hyphal complexes which assist in binding other hyphae together; they lack clamp connexions and are mostly thick-walled, but tend to be narrower than skeletal hyphae.

These hyphal types are combined into monomitic, dimitic or trimitic

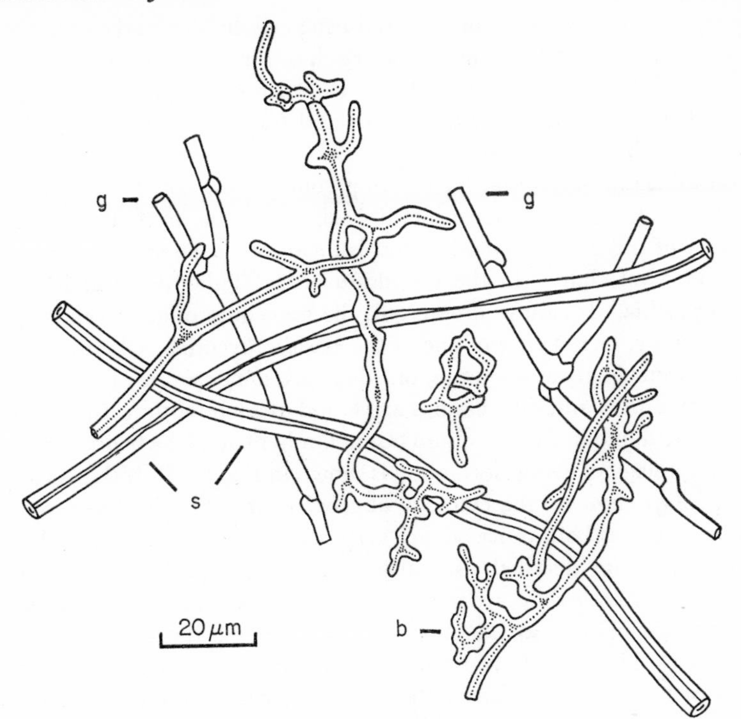

Fig. 70. *Coriolus hirsutus*, dissected hyphae. A trimitic hyphal system comprising generative hyphae with clamps (*g*), binding hyphae (*b*) and skeletal hyphae (*s*).

hyphal systems. In a **monomitic** system, only generative hyphae are present in the basidiocarp. A **dimitic** hyphal system combines generative hyphae with either skeletal or binding hyphae. A **trimitic** system combines all three hyphal types. Since the other types of hyphae are derived from generative hyphae it is to be expected that intermediates might occur, but although there is sometimes difficulty in interpreting which hyphal types and systems are present in a basidiocarp, these are, nevertheless, most valuable taxonomic characters.

The hyphal systems encountered among the Agaricales and 'Gasteromycetes' have not been thoroughly investigated, but do not readily fit the patterns described above, which apply mainly to the Aphyllophorales.

CLAMP CONNEXIONS

The hyphae of many species of Basidiomycotina have structures known as clamp connexions (Fig. 71). These are short appressed hyphal branches which originate at the same time as mitosis occurs in the dicaryotic nuclei, bypass the transverse septum formed during cell division and connect the two adjacent daughter cells of the hypha. Each originates from the distal cell and grows back to make contact and fuse with a minute 'peg' on the proximal cell. When viewed laterally the bypass and transverse septum together look not unlike a buckle or clamp.

Not all Basidiomycotina have clamp connexions; but if these are present in a mycelium it is an almost infallible sign that the mycelium belongs to a member of the Basidiomycotina and is dicaryotic. (There have been a very few reports of clamps in the mycelium of the Tuberales, but in no other Ascomycotina.) If clamps are absent, the mycelium might be monocaryotic or dicaryotic, or not that of one of the Basidiomycotina. Clamp connexions usually occur singly at a septum, but there are some species in which opposite paired clamps or multiple clamps are found at each septum. In some species clamps are found in the mycelium but not in hyphae of the fruitbody.

The cytology of clamp-formation is shown in Fig. 71, B. If this is compared with Fig. 53 showing the cytology of crozier-formation in Ascomycotina, it will be seen that both processes are very similar and occur only in dicaryotic hyphae. A more detailed comparison of the two processes is deferred to Chapter 13. Meanwhile, we may merely note that many mycologists regard croziers and clamps as homologous and, arguing that such similarity in cytology could hardly have been fortuitous or have arisen frequently in the course of evolution, they derive the Basidiomycotina from the Ascomycotina. They also consider that the basidium is homologous with an ascus, regarding the sterigmata and basidiospores as being enclosed by extensions of the wall of the basidium and thus comparable with the endogenous ascospores enclosed by an ascal wall.

Clamp connexions are apparently a device for ensuring that sister nuclei arising from conjugate division do not remain in the same daughter cell, but instead are segregated in such a way that the dicaryons are restored in the daughter cells. Especially with hyphae where width does not permit the mitotic spindles to lie side by side, this would be an advantage; but the fact is that equally narrow hyphae may have no clamps and yet appear to maintain their dicaryons without them. In some instances proliferation may take place from clamps. It has also been suggested that the presence of clamps facilitates the conduction of cytoplasm towards growing apices, since at each clamped septum there are two septal pores instead of only one between the adjacent cells.

REPRODUCTION AND FORMATION OF BASIDIOCARPS

Asexual reproduction in Basidiomycotina may take place by means of conidiospores, thallospores, and fragmentation. While asexual spores are not

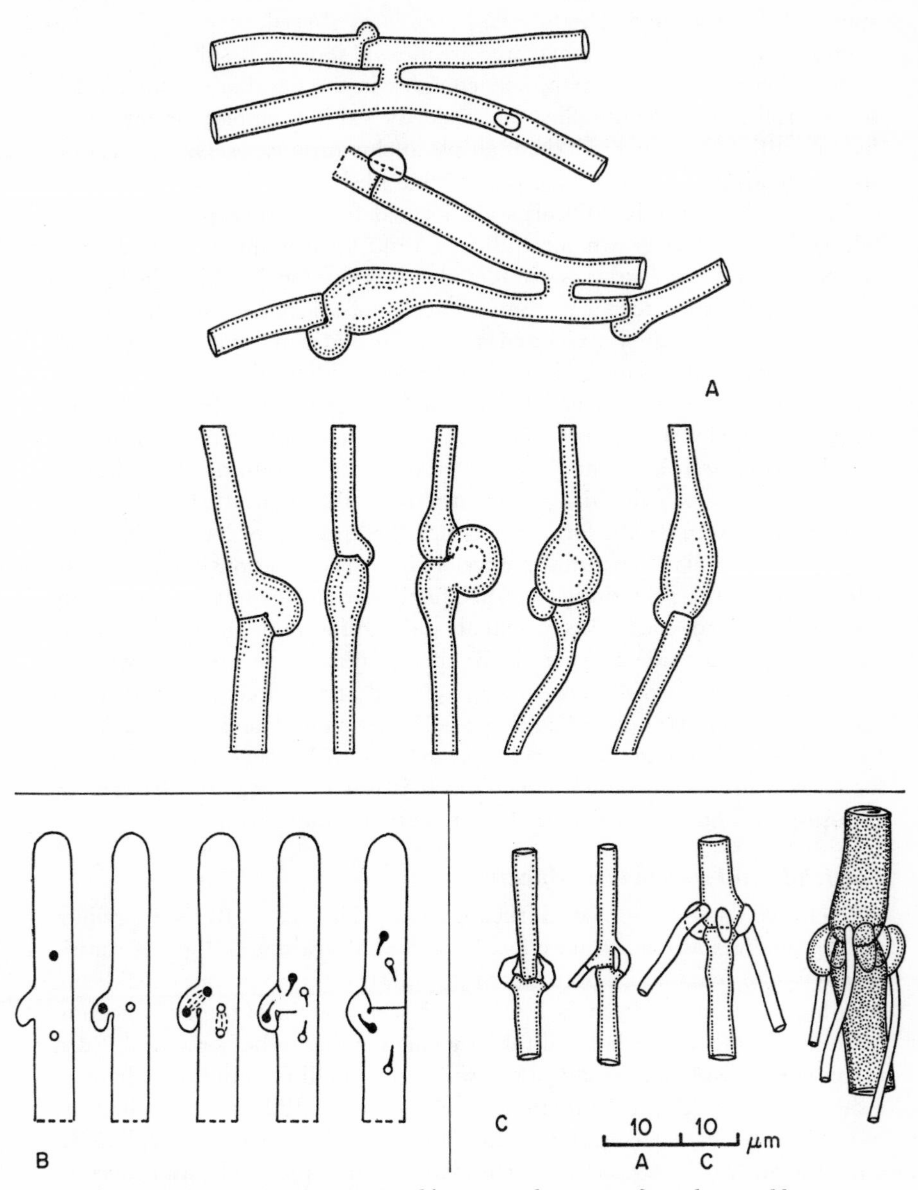

Fig. 71. Clamp connexions. A, Trechispore sp., clamps seen from above and laterally, some being swollen into 'ampoules'; B, cytology of clamp formation (diagrammatic); C, opposite paired clamps and multiple clamps (right) in *Coniophora puteana*; branch hyphae come from between the multiple clamps.

prominent in most Basidiomycotina they are important as the chief dispersal spores of the rust fungi (Uredinales) in the form of uredospores and aecidiospores. Some of the Deuteromycotina are asexual states of Basidiomycotina.

Sexual compatibility is extremely complex in the Basidiomycotina. While some species are homothallic, the majority exhibit bipolar or tetrapolar heterothallism, frequently of the multiple allelomorph types. Great variation in the genetic makeup of basidiospores is thus likely.

In most instances basidiocarps are formed from a dicaryotic mycelium which, after it has grown for sufficient time to accumulate the required reserves of food and energy, is stimulated to produce foci in which the hyphae branch prolifically to form compact tufts of plectenchymatous tissues. Further growth of these **primordia** (s. **primordium**) may be highly coordinated and subject to the interplay of various internal and external (e.g. light, gravity) form-factors which result in the production of fruitbodies with various characteristic shapes (Fig. 72). Some of the common forms are cupular (**cyphelloid**), club-like or coralloid (**clavarioid**), semicircular shelfs or brackets (**dimidiate**), or umbrella-like with stipe and pileus. Alternatively, some of the form-factors may be 'missing' or have little effect, and the primordia then usually develop into fructifications which spread laterally over the substratum as an effused mass of loosely or compactly arranged hyphae, indefinite in extent and outline and giving rise to the hymenium on the side away from the substratum. These fructifications are often termed **corticioid** when they are effused and entirely attached to the substratum, and **stereoid** when they are largely effused but reflexed towards the margin. These terms, however, apply to forms with a 'smooth' hymenium; for forms with a convoluted hymenium it is usual to speak of **effused** (or **resupinate**) and **effuso-reflexed** fructifications, respectively.

Corticioid and stereoid fructifications

In effused and effuso-reflexed fructifications, (Figs. 72, A, B) the mycelium usually emerges at one or more foci from the substratum and grows centrifugally while remaining appressed to the substratum, sometimes coalescing with hyphae from adjacent foci. Growth in the margin of the fructification is indeterminate and radial, with the hyphae tending to be horizontal. Even in this young state there is often a certain amount of differentiation of hyphae a short distance behind the margin to form sterile cystiidiform structures or, in dimitic species, skeletal hyphae in addition to the generative hyphae. In some species with very simple fructifications, the appressed basal layer of hyphae may give rise almost immediately to short vertical branches from which the basidia are formed. In others, the vertical hyphae grow more extensively to form a context of varying thickness before the hymenium is produced over its distal surface. The operation of form-factors is sometimes shown by the direction taken by cystidia with long pedicels: these may be

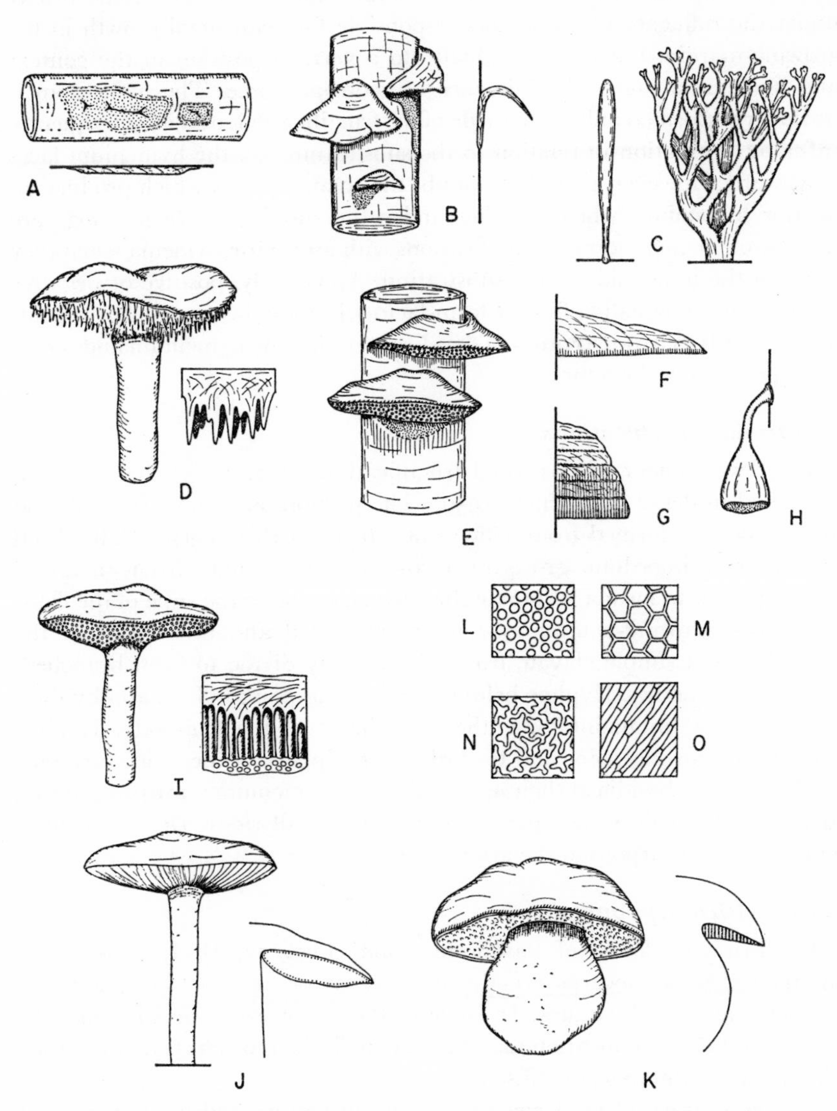

Fig. 72. Some characteristic fruitbodies of Basidiomycotina. A, effused (corticioid); B, effuso-reflexed (stereoid); C, clublike or coralloid (clavarioid); D, stipitate-pileate with dentate hymenophore (hydnoid); E, shelf-like (dimidiate); F, dimidiate-applanate; G, dimidiate-ungulate; H, cupulate (cyphelloid); I, stipitate-pileate with tubular hymenophore (polyporoid); J, stipitate-pileate with lamellate hymenophore (agaricoid); K, stipitate-pileate with tubular hymenophore and fleshy texture (boletoid); L, M, N, O, various shapes of pore mouths in polyporoid fungi.

horizontal in the basal layer of the fructification where they come briefly under the influence of the factors responsible for centrifugal growth in the growing-margin, but curve gradually to a vertical position in the context and hymenium under the influence of geotropic force. The hymenium is **unilateral** (formed only on one side of the fructification) and most frequently **inferior** in position in relation to the substratum, i.e. the hymenium faces downward. However, there is no doubt that many species which produce an inferior hymenium when growing under the substratum (e.g. a log) are capable of forming normal fructifications with **superior** hymenia when they grow on the upper side of the substratum. Apparently, positive or negative geotropism are equally efficient form-factors in these instances. These fruit-bodies are all **gymnocarpous** in development, i.e. the hymenium and spores are exposed from the outset.

Clavarioid fructifications

The development of clavarioid fructifications (Fig. 72, C) has been investigated in detail by Corner (1950). The primordia, which are small and subglobose, are formed from a mycelium or from rhizomorphs. The distal end of the primordium grows in a co-ordinated manner from an apical growing-point to form a turbinate shaft which becomes the stipe of the fruit-body. The growing-point may remain undivided and thus result in the formation of a simple clavate fruitbody, or may divide to form branched, coralloid fruitbodies. Hyphae behind the growing-point inflate, and by their turgidity assist in maintaining the erect habit of the fruitbody. Hyphae behind the growing-point and towards the periphery of the shaft, turn outwards and form basidia at their apices; thus the hymenium is **amphigenous,** covering the fertile upper part of the shaft on all sides. Development is typically gymnocarpous in clavarioid fungi.

Forms with a stipe and pileus

The principal forms of Basidiomycotina (Figs. 72, D, I, J, K) with umbrella-like basidiocarps are the mushrooms or agarics, the boleti, the polypores and the hydnums. In some of these, however, the stipe may be missing and the fructification may then be shell-like (**conchate**), fan-shaped (**flabellate**), dimidiate, or effused.

The basidiocarp in most agarics is fleshy in texture and its development is **hemiangiocarpous;** the hymenium forms over the surface of gills (**lamellae,** s. **lamella**) which are enclosed by basidiocarp tissue at first but later become exposed to the open air before the spores are mature. The primordium, even when only a few millimetres high, is already differentiated internally into distinct tissues comprising the stipe, pileus and lamellae, which expand into the adult mushroom over a short period, the expansion occurring mainly at night. During expansion the lower part of the pileus splits away from the stipe by a transverse circular split which exposes the lamellae.

A few agarics develop gymnocarpously, as do most boleti, polypores and hydnums. In these stipitate forms, the primordium has an apical growing-point which elongates to form the stipe, usually under the influence of positive phototropism and negative geotropism. After a time the apex ceases to elongate and spreads out centrifugally to form a diageotropic pileus. Towards the lower surface of the pileus the hyphae turn downward under the influence of positive geotropism, to form lamellae (in agarics), tubes (in boleti and polypores) or spines (in hydnums). The hymenium covers the surfaces of lamellae and spines, and lines the interior surfaces of tubes; it is thus usually inferior in position. The polypores and hydnums are typically woody, corky or membranous in texture and their tubes or spines are firmly united with the context tissues of the pileus. The boleti and agarics are typically fleshy in texture, while their tubes or lamellae are usually easily separable from the pileus.

Development in 'Gasteromycetes'

The 'Gasteromycetes' include Basidiomycotina whose basidiocarps are commonly known as puffballs, earthstars, or phalloids, as a reflection of their characteristic shapes (Figs. 81–85). Typically, these forms are **angiocarpous;** the fertile portion (**gleba**) of the fruitbody is enclosed in basidiocarp tissue until well after the spores have matured.

Cunningham (1944) has detailed four main types of development occurring in various 'Gasteromycetes'. In the majority of genera, distributed among four of the five recognized orders, development is of the lacunar type, e.g. *Rhizopogon*. In *R. rubescens* a mycelial strand gives rise to a small rounded or clavate primordium composed at first of loosely intertexed hyphae. Increased branching of hyphae within the primordium results in a compact tissue in which a few large schizogenous cavities appear, while the outer hyphae form a membranous covering, or peridium. The cavities are surrounded by compact plates of sterile tissue (**tramal plates**) which branch and anastomose to divide the cavities into a number of smaller ones. At length the hyphae of the tramal plates give rise to basidia which project into the cavities. Spores are soon formed and the basidia collapse. In some genera sheaths of 'nurse cells' envelop the immature spores after collapse of the basidia, and the spores continue to grow to maturity. In some genera the tramal plates remain, dividing the gleba into a number of chambers; in others they are digested. Finally, there may be little else within the peridium but dry powdery mature spores and perhaps remnants of hyphae forming a capillitium. Development of this type is common to most puffballs.

Another principal type of development is that found in the phalloids, e.g. *Phallus* (Fig. 83). The primordium is formed underground as an ovoid 'egg' as much as 5–6 cm in diameter, with a thin membranous peridium surrounding a subgelatinous mass in which the various parts of the fructification become differentiated in miniature. These consist principally of a spongy stipe (**receptacle**) with an apical dome of tissue, part of which forms the small pileus and part the basidiferous glebal tissue surrounding the pileus. After the spores are mature, the turgid spongy receptacle expands with sufficient force to rupture the peridium, which remains as a cup or **volva** at the base of the fructification.

HYMENIUM

Despite the fact that Ascomycotina and Basidiomycotina sometimes produce fruitbodies of rather similar external form—for instance a cup-like fruitbody — these differ in their orientation with respect to the substratum. In Ascomycotina spore liberation usually depends on a mechanism involving turgidity of the asci, and the fruitbodies face upward in a position such that they can hold moisture. In Basidiomycotina, however the spore liberation mechanism requires a humid atmosphere, but will not work when the basidia are wet. Almost all types of basidiocarps (except clavarioid ones and those of the 'Gasteromycetes') have the hymenium facing downward, a position in which the basidia are protected from free water but still have humid surroundings. The basidiospores are actively discharged to a distance of about 100–200 μm from the sterigmata. With an upward-facing hymenium the basidia would become wet, and any discharged basidiospores would readily fall back on the hymenium and fail to be dispersed. We have seen that, commonly, the basidiocarps are raised on stipes or take the form of brackets supported on substrata well above ground level; also that the hymenium is commonly spread over the face of lamellae, over spines, or within tubes. All these configurations of the hymenophore tissues greatly increase the spore-bearing area as compared with a smooth hymenium, without unduly increasing the bulk of the basidiocarp. The various forces that interact during the formation of a stipitate or bracket form of basidiocarp finally ensure that the lamellae, spines or tubes are very precisely positioned in a vertical plane. Spores shot into the spaces between these structures are thus free to fall vertically and be carried away by air currents without danger of their being intercepted by tissues of the basidiocarp. Ingold (1965) pointed out that most of the large fleshy Basidiomycotina are mesophytic; the atmosphere surrounding the hymenium is kept nearly at saturation point by moisture provided from the flesh of the fungus even when the ambient air has a low relative humidity. On the other hand, most gelatinous or leathery Basidiomycotina are regarded as drought-enduring zerophytes.

Donk (1964) has recognized certain categories of hymenia that are of taxonomic importance, particularly in the study of effused Aphyllophorales.

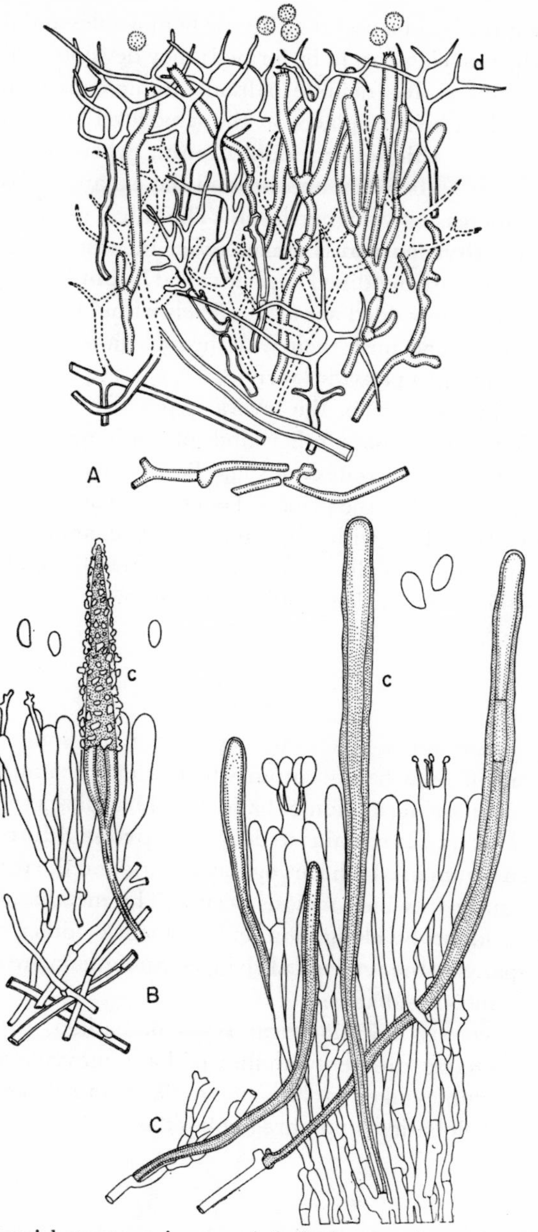

Fig. 73. Hymenial structures in some Aphyllophorales. A, catahymenium with dichohyphidia and basidia in *Scytinostroma portentosum*; B, hymenium of *Lopharia mirabilis* with basidia, a single large metuloid cystidium, and two dendrohyphidia; C, hymenium of *Columnocystis abietina* with basidia, and cystidia formed from skeletal hyphae; one cystidium (right) contains two adventitious septa; *c*, cystidium *d*, dichohyphidium.

In a **euhymenium,** basidia and their sterile homologues and derivatives are the first and principal elements to be formed; they develop at about the same time and are arranged in a definite palisade. Some euhymenia are **static,** the exhausted basidia being replaced at the same level by a process of intercalation. Others are **thickening euhymenia,** where branches of subhymenial hyphae grow between and beyond the exhausted basidia to form new basidia at increasingly higher levels. The other principal type of hymenium is the **catahymenium** (Lemke, 1964) in which the first elements to be formed are variously modified hyphae called **hyphidia** (s. **hyphidium;** Fig. 73); basidia are embedded at various levels within the mass of hyphidia and have to elongate and push their way through these to the surface. The basidia thus do not form a palisade layer.

Instead of, or in addition to, the several types of hyphidia that may be present in the hymenium, many other kinds of sterile modified hyphae may occur in the hymenium or context of some Basidiomycotina. These are too numerous to detail here, but the principal ones are an assortment of **cystidia** (Fig. 73), **gloeocystidia** (with oily contents) and **setae.** The reader is referred to Lentz (1954), Talbot (1954), Donk (1964), Lemke (1964) and A. H. Smith (1966) for discussions of these structures.

BASIDIA

The typical basidium is a clavate, cylindrical or somewhat rounded terminal cell cut off by a basal septum which is often clamped. The main cytological events in the formation of basidia and basidiospores have already been mentioned (p. 97). It remains to define the **probasidium** as that part or stage of the basidium in which karyogamy occurs, and the **metabasidium** as that part or stage in which meiosis occurs. The metabasidium develops out of the probasidium by elongation, enlargement or other morphological changes accompanying the cytological changes. Sterigmata are formed as outgrowths from the metabasidial wall.

There are numerous morphological types of basidia, some distinctive enough to characterize different families of Basidiomycotina. Two main, general types, sometimes taken to characterize classes or subclasses, must now be distinguished: holobasidia and phragmobasidia.

Holobasidium

A holobasidium (Fig. 74) is a basidium whose metabasidium is not divided by primary septa. It is typically clavate or cylindrical, with four relatively short, uninflated, curved sterigmata (rarely fewer than four, or up to eight sterigmata). Typically, the basidiospores germinate by a germtube to form hyphae directly. The chief exceptions occur in the Dacrymycetales and

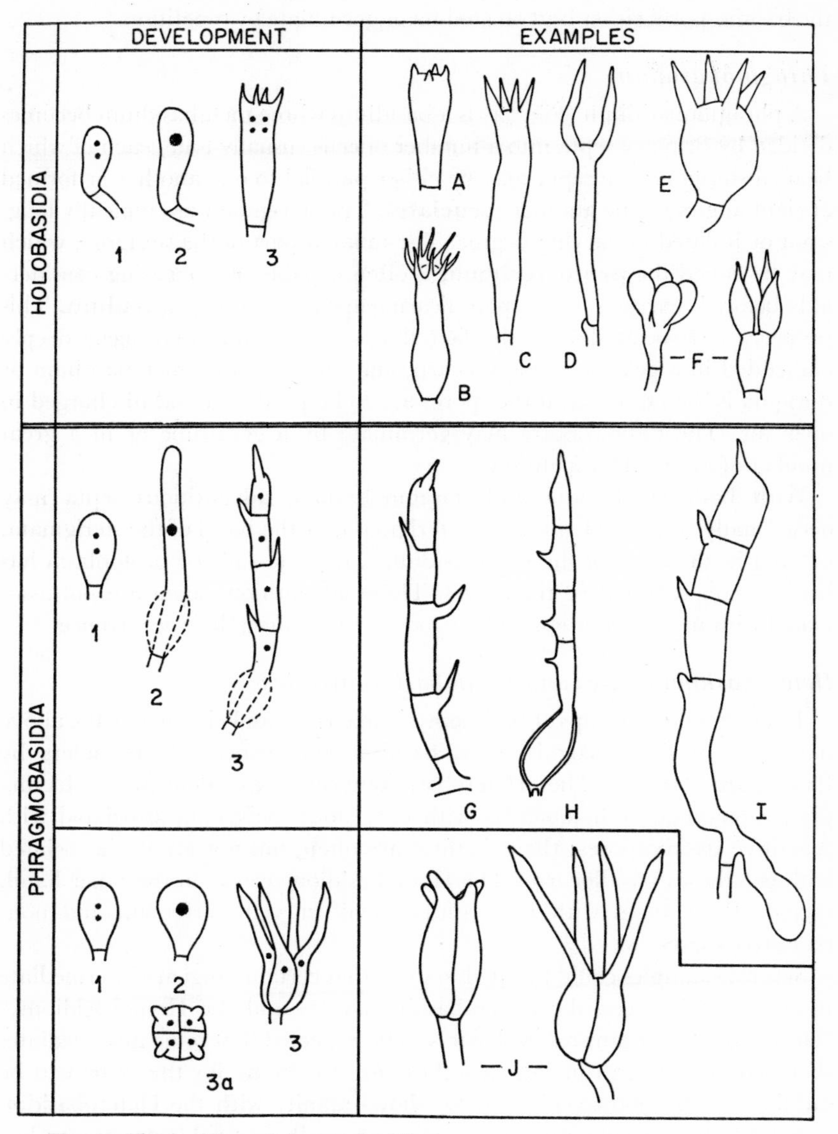

Fig. 74. Holobasidia and phragmobasidia.
Development: 1, probasidium; 2, karyogamy; 3, metabasidium and sterigmata; 3*a*, metabasidium viewed from above.
Examples: (camera lucida drawings at different magnifications), A, *Agaricus*; B, *Sistotrema*; C, *Aleurodiscus*; D, *Dacrymyces*; E, *Ceratobasidium*; F, *Tulasnella*; G, *Helicobasidium*; H, *Septobasidium*; I, *Helicogloea*; J, *Tremella*.

Tulasnellales where the holobasidia have strongly inflated sterigmata and the basidiospores either bud off conidia or germinate by repetition.

Phragmobasidium

A phragmobasidium (Fig. 74) is a basidium whose metabasidium becomes divided by primary septa into a number of cells (usually four) each of which bears a sterigma; the septa may be either parallel to one another or formed at right angles to one another (**cruciate**). The sterigmata are typically long, stout or inflated in varying degree. The inflated part of the sterigma, which may be called the **protosterigma,** is often capable of elongating considerably before it narrows to the spore-bearing apex termed the **spiculum.** This property is associated with the fact that such basidia often occur deeply embedded in a gelatinous basidiocarp, and an elongating metabasidium or sterigma is then essential if the spores are to be produced and discharged in open air. The basidiospores may germinate by a germtube or in a great number of species, by repetition.

With both holobasidia and phragmobasidia, adventitious septa may occasionally be formed within the sterigmata, at the bases of the sterigmata, or even in the body of the metabasidium when most of the protoplasm has become concentrated in the spores. These adventitious septa are not associated with nuclear division and are mostly irregular in their occurrence.

Heterobasidiomycetes and Homobasidiomycetes

Two taxonomic groups of Basidiomycotina, commonly known by the above names, were first recognized on basidial and associated characters by Patouillard (1900). The Heterobasidiomycetes are defined as having phragmobasidia, or holobasidia with very stout sterigmata associated with repetitive basidiospores; these features are often, but not always, associated with gelatinous basidiocarps. The Homobasidiomycetes, on the other hand, include those types with holobasidia, small curved sterigmata and non-repetitive spores.

Several examples could be cited to show that certain fungi are intermediate in character between the Heterobasidiomycetes and the Homobasidiomycetes; and since organisms evolve it is to be expected that some intermediates should occur; though in this case there are too many for the system to be satisfactory. *Dacrymyces* (Fig. 74, D) shows affinity with the Heterobasidiomycetes by having gelatinous fruitbodies and swollen, variable sterigmata, but with the Homobasidiomycetes by having holobasidia and non-repetitive basidiospores (which often bud off conidia). *Ceratobasidium* (Fig. 74, E) is allied to the Heterobasidiomycetes by having stout, variable sterigmata and repetitive basidiospores, but to the Homobasidiomycetes by having holobasidia and non-gelatinous basidiocarps. Because of difficulties such as these, Talbot (1965, 1968) proposed that the groups Heterobasidiomycetes and

Homobasidiomycetes should no longer be recognized, but stressed the great usefulness of all these characters in defining lower taxa. The classification adopted in this book omits recognition of these classes, and of another which was proposed (Lowy, 1968) to accommodate the intermediates.

Basidial terminology

Since phragmobasidia, particularly, are not only diverse in type but also tend to be rather variable within one type, several views have arisen on the homology of basidial parts and thus on the terms to be applied to these. The terminology still used by most mycologists is that proposed by Neuhoff (1924) and supported by Rogers (1934) and Martin (1938). In holobasidia, Neuhoff referred to the body of the basidium as the **basidium** and the spore-bearing extensions as **sterigmata.** But in phragmobasidia he was impressed by the large and variable spore-bearing extensions which can elongate to meet the ecological situation and finally to produce a narrowed spore-bearing apex on reaching the outer surface of the fruitbody. He termed the body of a phragmobasidium a **hypobasidium** and the extensions, **epibasidia;** the narrowed apices of the epibasidia he considered to be the true **sterigmata.**

The main defects in Neuhoff's terminology were first exposed by Donk (1931, 1954), followed by Talbot (1954). First, compare the basidia of *Agaricus*, *Tremella* and *Septobasidium* in Fig. 75, where the parts are labelled according to Neuhoff's terminology. Note that the shaded parts in the *Septobasidium* type of basidium correspond exactly with the epibasidia of *Tremella*, but received no name; the epibasidium of *Tremella* is clearly not homologous with that of *Septobasidium*. Second, genera such as *Ceratobasidium* and *Dacrymyces* have basidial characters intermediate between those characterizing the Homobasidiomycetes and Heterobasidiomycetes, and have, in fact, been classed at different times in both these groups. With Neuhoff's terminology, if these genera were classed as Heterobasidiomycetes the spore-bearing parts would be called epibasidia; but they would be called sterigmata if the genera were classed as Homobasidiomycetes. The sterigmata themselves do not change with change in classification, and they should always receive the same name. These and other criticisms led Martin (1957) to re-define some of Neuhoff's terms in a new way, and he also adopted the term **metabasidium** in a new context. It is apparent from this paper that there is now substantial agreement between Martin's and Donk's views on homology, though their terms differ; this fact does not seem to have been appreciated, however, by many mycologists who continue to use Neuhoff's terms unchanged.

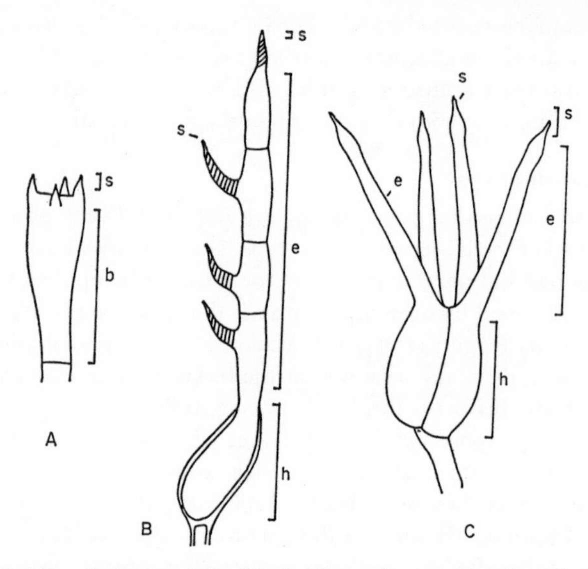

Fig. 75. Basidia of *Agaricus* (A), *Septobasidium* (B) and *Tremella* (C), labelled according to Neuhoff's terminology; *b*, basidium; *e*, epibasidium; *h*, hypobasidium; *s*, true sterigma. The shaded parts received no name.

DISCHARGE OF BASIDIOSPORES

With basidia that discharge their spores forcibly, the spore is borne obliquely at the tip of the sterigma and, as it becomes mature, a lateral bubble is formed next to the **hilum** ('apiculus') at the base of the spore. The exact nature of the bubble has long been a puzzle. Buller (1909, 1922) thought that it was exuded water, and noted that it remains on the spore after the latter is discharged. Olive (1964) suggested that it might be a bubble of carbon dioxide, which enlarges and suddenly bursts with enough force to throw the spore a short distance from the sterigma. Savile (1965) agreed that bursting of such a bubble would provide the energy necessary for fracturing the line of attachment to the sterigma and releasing the spore, but pointed out that the force is not applied in an appropriate direction to throw the spore away from the hymenium in the axis of the sterigma; he suggested that like electrostatic charges on the spore and sterigma would provide the repulsive force necessary to discharge the spore once its attachment had been broken. Electron microscopy (Wells, 1965) has shown that the bubble occupies a space between the spore wall and an inflated extension of the wall of the sterigma, and that the bubble accompanies the spore on discharge. Wells postulates that the contents of the bubble are unlikely to be gaseous, but instead must have high osmotic value and thereby draw water from the sterigma or the spore or the environment, until bursting occurs. The attachment may be weakened by enzymatic processes and simultaneously by the tendency of the apex of the sterigma to round off under turgor pressure from the metabasidium.

Basidiospores forcibly discharged by the 'drop-excretion mechanism' are often termed **ballistospores;** once they have been launched into the air they are free to be dispersed by air currents. Although ballistospores are characteristic of those Basidiomycotina with exposed hymenia, forcible discharge would be of no advantage and apparently does not occur in those angiocarpous Basidiomycotina with hymenia enclosed until after the spores are mature, i.e. the 'Gasteromycetes'. The sterigmata in these types are either very long and thin, or very small and spicular; in either case they are readily fractured when the basidia collapse or as the spores reach maturity. The spores then lie free to be dispersed passively by insects or water (with slime-spores) or by wind, splashes of rain or mechanical disturbance (with dry spores).

The majority of basidiospores germinate by means of a germtube to a form mycelium directly, but germination by repetition is common among the Basidiomycotina with phragmobasidia. In a few groups (e.g. Dacrymycetales and Ustilaginales) the basidiospores may bud off blastospores which then anastomose or germinate by a germtube.

Very few basidiospores become septate; the majority are relatively small, colourless and smooth, though a smaller number may be coloured and/or ornamented with small spines or warts. Some basidiospores give a positive **amyloid** reaction with Melzer's reagent, a feature which is put to taxonomic use though with variable emphasis. The capacity of some thick-walled spores to adsorb aniline blue dye strongly to the inner-wall surface gave the clue to relationship among a number of genera which had previously been scattered in several different families, and led to the establishment of a natural family, the Coniophoraceae, for these genera. Such spores are said to be **cyanophilous.**

TAXONOMIC IMPLICATIONS

As might be expected with fungi that mostly form large basidiocarps, the earlier classifications of Basidiomycotina were based mainly on gross characters visible with the naked eye or handlens. The broad classification derived from the works of Fries culminating in *Hymenomycetes Europaei* (Fries, 1874) is set out simply in Saccardo's *Sylloge Fungorum* (1887, 1888). The main features used for separating taxa were the development of the basidiocarp (gymnocarpous or hemiangiocarpous *versus* angiocarpous), and the position of the hymenium in relation to the substratum, and the convolution of the hymenophore. Subsequently, Patouillard (1900) stressed the nature of the basidium and on this basis recognized the classes Heterobasidiomycetes and

Homobasidiomycetes. These classifications can be partly summarized as follows:

Friesian Classification and Homoplasy

Class Heterobasidiomycetes (p. 206)

Orders Uredinales and Ustilaginales: plant parasites with sori containing teleutospores (which function as probasidia).

Order Tremellales: mostly saprobic with well developed, often gelatinous basidiocarps.

Class Homobasidiomycetes (p. 206)

Series Hymenomycetes: developing gymnocarpously or hemiangiocarpously.

Order Aphyllophorales: with a smooth, dentate, or tubular hymenophore (if tubular associated with basidiocarps of woody, corky or membranous texture).

Family Clavariaceae: hymenium amphigenous; fruit-bodies erect, club-like or branched.

Family Thelephoraceae: hymenium smooth, inferior; fruitbodies effused, effuso-reflexed or flabellate.

Family Cyphellaceae: hymenium smooth, inferior; fruitbodies cupular or of discrete tubules.

Family Hydnaceae: hymenium spread over dentate processes, inferior.

Family Polyporaceae: hymenium inferior, lining the inside of tubes or pores, not easily separable from the context tissue of the fruitbody.

Order Agaricales: hymenium spread over lamellae or lining tubes, inferior (if tubular, associated with fleshy basidiocarps and the tubes easily separable from the context tissue).

Family Agaricaceae: hymenium spread over lamellae.

Family Boletaceae: hymenium lining tubes.

Series Gasteromycetes: developing angiocarpously.

Order Lycoperdales: epigean; the gleba powdery at maturity.

Order Phallales: epigean; basidiocarp fleshy-gelatinous, erupting from a volva.

Order Nidulariales: epigean; the gleba sphaeroid or divided into several lenticular parts.

Order Hymenogastrales: hypogean.

It is possible to suggest a plausible evolutionary sequence running through the forms recognized in the above classification. One can observe a sequence in which there is increasing protection of the hymenium until the spores are ripe; a progression to forms with a protected inferior hymenium where the basidiospore discharge mechanism will work most efficiently; and a sequence in which the hymenium becomes compacted and convoluted, thus progressively increasing the spore-bearing area without wastefully increasing the size of the basidiocarp. All these are undoubtedly evolutionary tendencies, but it is a mistake to express them in a rigid series without being sure of the starting-point and direction of evolution, and especially without assessing the extent to which parallel evolution may have occurred. There is a great danger of mistaking homoplastic forms arrived at by parallel evolution, for those which are truly homologous.

There is now abundant evidence that the Friesian classification is highly artificial, and that series of closely related species or genera cut right across the lines of Fries's families, orders or classes. Some examples will now be considered.

Corner (1950) found at least six microscopic ways in which the clavarioid basidiocarp can be constructed. Many of the genera are not at all closely related and have to be removed from a restricted family Clavariaceae. The genera *Lachnocladium* and *Clavariachaete* show the artificiality of the Clavariaceae, since they fit closely with a series of genera drawn from the Polyporaceae (*Fomes* and some other polypores), the Hydnaceae (*Asterodon*), the Thelephoraceae (*Hymenochaete*, *Asterostroma*), and even with the lamellate *Cyclomyces*. These genera are now sometimes taken to constitute a family, Hymenochaetaceae, or a series known as the Xanthochroic Series. This taxon picks species from their old genera, genera from the old families, and relates them by their uninflated hyphae, their dimitic hyphal systems, their pigmentation, the frequent presence of setae in the hymenium, and their colour reaction with alkali.

The Thelephoraceae is one of the most artificial of Fries's families, whose members merge especially with the Hydnaceae. According to Corner (1954) the resupinate forms are derived from bracket and pileate forms from which the form-factors have disappeared; thus they are a mixture of end-lines of many Basidiomycotina that have reached this stage of efficiency. This is quite the reverse of the usual interpretation put on the Thelephoraceae, as most workers have hitherto assumed that the higher Hymenomycetes must have been derived from the simpler resupinate fungi. Two series or families with members derived largely from the resupinate Thelephoraceae are the *Coniophora* Series (or Coniophoraceae) and the *Thelephora* Series (Thelephoraceae in a restricted sense). The Coniophoraceae draws *Coniophora*, *Jaapia* and *Serpula* from the old Thelephoraceae; *Meruliporia* from the Polyporaceae; and *Gyrodontium* and *Amaurodon* from the Hydnaceae. As well as having cyanophilous spores, members of these genera have a number of other microscopic features in common, which outweigh their external dissimilarities. The restricted family Thelephoraceae includes *Thelephora* and *Tomentella* drawn from the old Thelephoraceae; *Caldesiella* (Hydnaceae); *Aphelaria* and *Scytinopogon* (Clavariaceae).

Among the Polyporaceae, a single species such as *Daedalea trabea* may have a hymenium which is poroid, lamellate or somewhat dentate, besides having a wide range of colour. Generic segregations based on such features are bound to fail. It is small wonder that, at least until recently, one usually learnt to recognize the species first and then the genera in which they were traditionally placed.

In the conventional classification of Agaricaceae there were six tribes based on the massed colour of the spores, and within these tribes the macromorphology of the volva, annulus, stipe, gill attachment and texture gave the genera. Singer's (1962) monumental work on the Agaricales has brought about the recognition of sixteen families in place of the single family Agaricaceae, by a masterly synthesis of macroscopic and microscopic features and by observation of combinations of features which indicate affinity, rather than by reliance upon one or two obvious, but relatively trivial, features.

The classical concept of the Agaricaceae and Gasteromycetes as distinct phyletic units is undergoing drastic revision. Heim (1948) drew attention to a very natural series, the Asterosporic Series (or Asterosporales), comprising genera drawn from both Agaricales and Gasteromycetes. The clue to their affinity was first found in their spherical basidiospores with amyloid ornamentation in many species, and in the presence of laticiferous hyphae and sphaerocysts in the context tissue. Sphaerocysts (Fig. 80, A) are rounded, inflated cells which give the context a spongy texture. The genera *Russula* and *Lactarius* (Russulaceae) include both gymnocarpous and pseudoangiocarpous forms; they are related to *Hydnangium*, *Octaviana* and *Secotium*, which are hypogean or subepigean members of the Hymenogastraceae (Gasteromycetes). In all these genera there are undoubted links in the method of development of the fruitbody, in pigmentation and in characters of the spores. In the Hymenogastraceae there are incomplete lamellar folds lining the gleba, and the gap between these and ordinary lamellae has been bridged in many recently discovered species of tropical mushrooms. Some of the black-spored, ochre-spored and pink-spored Agaricaceae have also been found to have counterparts among the Gasteromycetes.

There are also a number of Agaricaceae in which the hymenia are poroid rather than lamellate. *Mycena* and *Marasmius* have true lamellae but are connected through a series of many tropical forms to species in which the hymenium has anastomosed veins. The more agaricoid of these have elongated lamellae with only a few transverse veins, but others become more alveolar and finally, at the end of the series, we have species with small tubular pores. *Favolaschia* and *Campanella* are examples of agarics with pores, and they are considered to be related to *Mycena* and *Marasmius* rather than to the Polyporaceae. Confirmation of such a relationship is seen in the presence of peculiar cells (acanthohyphidial cells) on the surface of the pileus in each.

These are but a few of the difficulties that have arisen in trying to classify the Basidiomycotina in a natural system, and of the unsatisfactory classifications resulting when the phenomenon of homoplasy is not fully appreciated. Classification is not static and could only become so if a perfect system were attained. One might well ask, what advantage is there in striving for a natural system when an artificial one, besides being easier to comprehend, can be used to classify a specimen with the least possible trouble? One difficulty is that specimens in different stages of development may have to be classified differently; even in a single specimen some parts may be typical of one genus and family while others may be typical of another genus and family. In such circumstances it is virtually impossible to match the specimen to a species, to file it away in the herbarium without elaborate cross-references, or to predict some of its properties from a knowledge of its apparent relatives. An artificial classification may seem satisfactory but often breaks down at the crucial moment. The Friesian system has obvious advantages in ease of comprehension and application in the field, but suffers from the disadvantage of being artificial. It is gradually being replaced by more sophisticated classifications used by specialists in certain groups of Basidiomycotina. Because the taxa need to be defined by a combination of many characters, especially the more stable microscopic ones, such classifications are infinitely more difficult to grasp and apply.

Key to the Classes and Orders of Basidiomycotina

1. Basidiocarp represented by a sorus of teleutospores (resting spores homologous with probasidia); parasites mostly of vascular plants — Class *Teliomycetes* 4.
   Basidiocarp usually well developed; basidia typically in a hymenium in or on basidiocarp tissue, less often in a limited hymenium arising in tufts or not on basidiocarp tissue; mostly saprobic — 2.

2. Mostly gymnocarpous or hemiangiocarpous with the hymenium exposed before the spores are mature; basidiospores mostly forcibly discharged, i.e. ballistospores (‘*Hymenomycetes*’) — 3.
   Mostly angiocarpous; basidiospores not ballistospores (‘*Gasteromycetes*’) — 9.

3. With phragmobasidia, regularly divided by primary septa in the body of the metabasidium; (often with a gelatinous basidiocarp; often with repetitive spores; usually with stout, variable sterigmata) — Class *Phragmobasidiomycetes* 5.
   With holobasidia, sometimes divided by adventitious septa especially in the sterigmata; (rarely a gelatinous basidiocarp; rarely with repetitive spores or stout variable sterigmata) — Class *Holobasidiomycetes* 6.

4. *Teliomycetes*
   Teleutospores terminal, giving rise to a horizontally septate metabasidium bearing forcibly discharged basidiospores on sterigmata — *Uredinales*
   Teleutospores mostly intercalary, giving rise to simple or branching, aseptate or horizontally septate metabasidia bearing lateral or terminal sessile basidiospores which are not forcibly discharged and frequently bud to form pseudomycelium — *Ustilaginales*

5. *Phragmobasidiomycetes*
   Metabasidium broad, divided cruciately by three longitudinal or sometimes oblique septa into four cells — *Tremellales*
   Metabasidium more or less cylindrical, divided horizontally by septa into one–four cells; basidiocarp not formed in association with scale insects — *Auriculariales*
   Metabasidium as in the Auriculariales but basidiocarps always associated with scale insects — *Septobasidiales*

6. *Holobasidiomycetes*
   Basidia cylindric-clavate, always with two stout sterigmata; basidiospores

frequently septate, frequently budding off small blastospores; basidiocarps mostly gelatinous *Dacrymycetales*

Basidia subspherical to broad, with a variable number (usually four) of stout finger-like sterigmata and repetitive basidiospores; or sterigmata inflated, cut off by basal adventitious septa and deciduous, with repetitive spores *Tulasnellales*

Basidia with small, uninflated sterigmata, or with long filamentous sterigmata, or with sessile basidiospores; spores not repetitive 7.

7. Internal parasites of leaves of higher plants
   (*a*) Basidia emerging in tufts from stomata; probasidium and cylindrical metabasidium separated by a constriction at maturity; basidia bearing two spores *Brachybasidiales*
   (*b*) Basidia emerging through the epidermis to form a hymenium of indefinite extent; basidia clavate, not constricted *Exobasidiales*

   Mostly saprobic; if an internal parasite then not as above but forming a distinct external fruitbody 8.

8. Basidiospores forcibly discharged (ballistospores); hymenium exposed before the spores are mature
   (*a*) Fruitbody developing basically gymnocarpously; hymenium unilateral or amphigenous, smooth, or covering dentate processes or lining cups or tubes; if tubular, the tubes may be discrete or coalesced but are firmly united with the tissue of the woody, corky or membranous basidiocarp *Aphyllophorales*
   (*b*) Fruitbody developing basically hemiangiocarpously or gymnocarpously; hymenium covering lamellae on the lower surface of the pileus, or if lining tubes then these are fleshy and easily separable from the fleshy pileus *Agaricales*

   Basidiospores not ballistospores. Spores produced in a hymenium in a closed fruitbody and not exposed until after maturity, i.e. basically angiocarpous ('*Gasteromycetes*') 9.

9. '*Gasteromycetes*'
   (*a*) Gleba fleshy, compact, indehiscent, composed of tramal plates anastomosed to form cavities lined with hymenium; lacking a capillitium; often hypogean *Hymenogastrales*
   (*b*) Gleba becoming mucilaginous and foetid, eventually exposed on a specialized spongy receptacle arising from a volva *Phallales*
   (*c*) Gleba powdery at maturity; lacking a capillitium *Sclerodermatales*
   (*d*) Gleba powdery at maturity; with a capillitium *Lycoperdales*
   (*e*) Glebal chambers forming peridiola carried in funnel-shaped peridia, or the entire gleba forming a single peridiolum carried in and ejected from a globose peridium *Nidulariales*

CLASS TELIOMYCETES

Order Uredinales

The Uredinales or rust fungi are of great economic importance as pathogens of higher plants, particularly grasses and cereals. Usually, they have been regarded as obligate parasites but Williams, Scott and Kuhl (1966) grew a strain of *Puccinia graminis f.sp. tritici* and succeeded in producing its teleutospores and uredospores in culture. The Uredinales produce no basidiocarps but form various kinds of spores in sori, and spermatia in spermogonia. Many Uredinales are restricted in host range but, because they are heterothallic, genetic recombinations are responsible for the existence of species in scores of genetically different strains with differing pathogenic qualities. *Puccinia graminis* is able to attack several species of cereals and grasses, but the *forma specialis* attacking wheat, for example, exists in scores of strains with varying capabilities for attacking different races of wheat. Control of rust disease in wheat is therefore directed largely at breeding races of wheat with resistance to known strains of *P. graminis*.

The mycelium of rusts is intercellular in the host plant and the hyphae produce haustoria which penetrate the host cells. The hyphae have uninucleate cells in some stages of the life-cycle but eventually become dicaryotic and give rise to thick-walled resting-spores (**teleutospores;** Fig. 76) which are dicaryotic and are formed in acervulus-like **teleutosori.** The teleutospores are homologous with probasidia; each cell of the teleutospore is capable of germinating (Fig. 77) into a **metabasidium** (sometimes called a promycelium) in which karyogamy and meiosis occur. The metabasidium becomes transversely divided into four uninucleate haploid cells, each of which bears a short sterigma. The nuclei migrate through the sterigmata to the developing basidiospores, each of which is haploid. In form, the basidium is an auricularious phragmobasidium and the Uredinales might thus be united with the Auriculariales; however, the absence of clamp connexions and dolipore septa suggests that the Uredinales should be kept separate. The teleutospores are resistant, and in cold climates usually represent the overwintering stage; in warm climates they are often absent from the life-cycle, or rare, and in some instances their functions may be assumed by another type of spore, the aecidiospore. Teleutospores may be one-celled to several-celled, sessile or pedicellate, free or aggregated in various ways into layers, crusts or columns; these features are widely used for classification of Uredinales into families and genera.

Life-cycle patterns

The Uredinales are pleomorphic, as many as five spore-forms being possible in the life-cycle of some species. In any one species the different spore-forms are formed in a definite sequence, but in different species there may be quite different cycles according to whether some of the spore-forms are omitted or modified. Various terms have been given to denote cycles of different pattern; e.g. a **macrocyclic** rust, produces all five consecutive spore-forms. With regard to their host-relationships, some Uredinales are **autoecious,** completing their life-cycle on one species of host, while others are **heteroecious** and inhabit different host species at different stages in the life-cycle.

Puccinia graminis will now be considered as an example of a typical

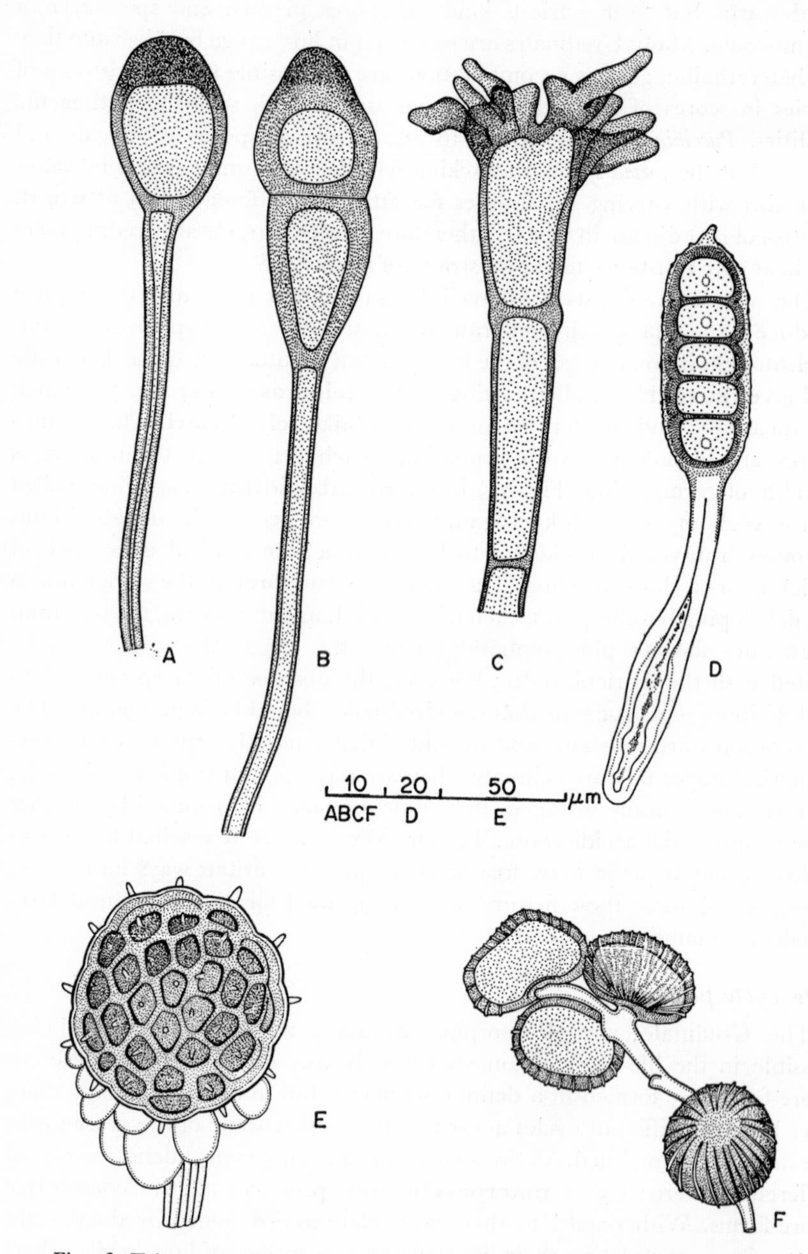

Fig. 76. Teleutospores: A, *Uromyces* sp.; B, *Puccinia graminis*; C, *Puccinia coronata*; D, *Phragmidium mucronatum*; E, *Ravenelia* sp.; F, *Uromycladium tepperianum*.

macrocyclic, heteroecious rust. The spore-forms are assigned a Roman numeral, from O to IV, as indicated below. In *P. graminis* stages II, III, and IV occur on the **primary host** (defined as the host bearing the teleutospore-stage) which is a member of the Gramineae, while stages O and I occur on the **alternate host,** *Berberis* (barberry).

O. Spermogonia with spermatia and receptive hyphae: These correspond to sex organs in the Uredinales. The spermogonia (Fig. 77), formed on primary haploid mycelium, are flask-shaped structures like pycnidia, immersed in the host tissue except for a protruding ostiole. Their interior is lined with **spermatiophores** bearing uninucleate haploid **spermatia** which have a male function and exude from the ostiole in droplets of nectar. The **receptive hyphae** grow out from the upper part of the spermogonium through the ostiole and into the drop of nectar. However, spermatia and receptive hyphae from the same mycelium are not compatible. Spermatization between two compatible mycelia is carried out by the agency of

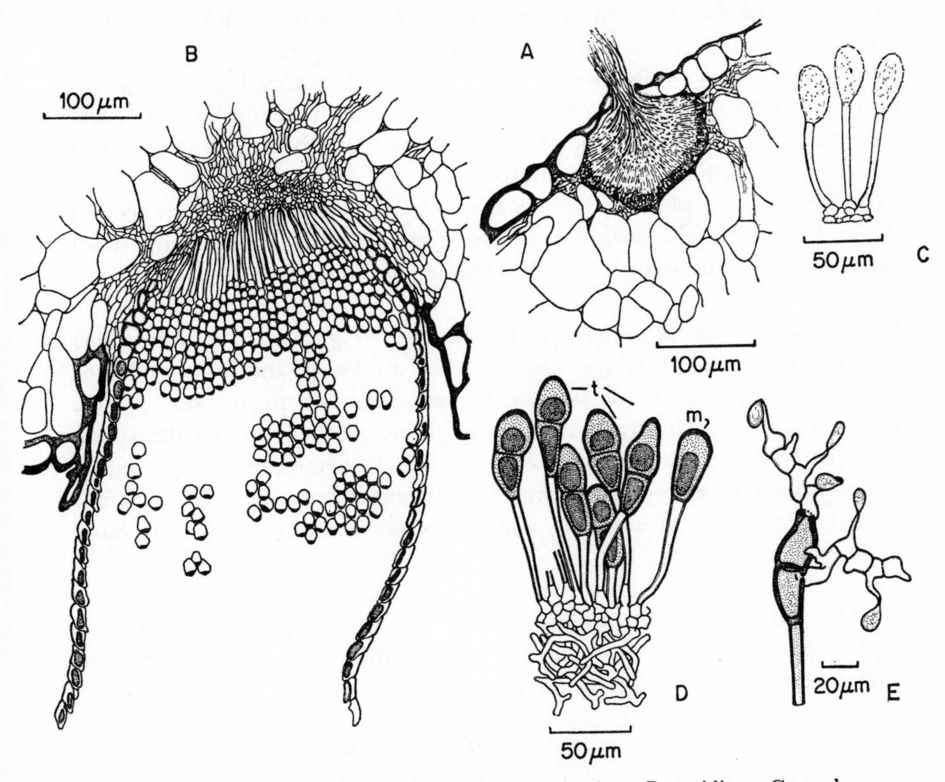

Fig. 77. Life-cycle of *Puccinia graminis*: A, spermogonium; B, aecidium; C, uredospores; D, teleutospores (t) and a mesospore (m); E, *Puccinia malvacearum*, germinating teleutospores showing metabasidia and basidiospores.

insects and results in a dicaryotic mycelium which ramifies through the barberry leaf and gives rise to aecidia.

I. Aecidia and aecidiospores: The aecidium (Fig. 77) is a cup-shaped structure partly immersed in the host tissue on the lower side of the leaves (or sometimes on petioles or stems). From the dicaryotic mycelium, sporogenous cells are formed at the base of the aecidium and by successive conjugate divisions of the nuclei, give rise to chains of unicellular binucleate aecidiospores separated by small disjunctor pads. Peripheral cells at the base of the aecidium become differentiated as a peridial wall, which at first covers the aecidium but later ruptures and forms the distal part of the cupulate structure, standing out perhaps a couple of millimetres above the level of the host surface.

II. Uredosori and uredospores: The uredosori, which resemble acervuli in at first being embedded but later erupting through the host epidermis, are formed from a dicaryotic mycelium originating from the germination of an aecidiospore or of another uredospore (Fig. 77). Uredospores are the chief propagative spores of the fungus, functioning as the asexual summer spores in cold climates but are often the predominant spore-type all the year round in warm ones. They are unicellular, ornamented with small spines, and germinate into a dicaryotic mycelium.

In the uredosori of many rusts, a fungal hyperparasite sometimes occurs; this is *Darluca filum*, a member of the Sphaeropsidales which forms small black pycnidia containing two-celled, biapiculate conidia.

III. Teleutosori and teleutospores: In *Puccinia graminis*, the teleutosori have essentially the same appearance as the uredosori, and in fact teleutospores (Fig. 77) and uredospores may sometimes occupy the same sorus. The teleutosori arise from a dicaryotic mycelium and the spores are dicaryotic at an early stage. **Mesospores** are one-celled teleutospores occurring in species which normally have two-celled teleutospores, as in *P. graminis*. Germination of the teleutospore and formation of the metabasidium have already been considered.

IV. Metabasidium and basidiospores: The basidiospores (Fig. 77) are haploid and, since rust mycelia are heterothallic, carry different genetic strain factors. They germinate and infect barberry.

Classification of Uredinales

Classification in this order is based primarily on the nature of the teleutosori and teleutospores; thus, if these are absent, determination of specimens may be difficult. Rust species can usually be determined only after the host has been identified and use made of a host index listing the species recorded on particular hosts. There is also much difficulty in connecting the various states of heteroecious rusts, i.e. in associating as one species the states found on quite different primary and alternate hosts.

In the Pucciniaceae and Melampsoraceae, the teleutospores develop an external auricularious metabasidium during germination, while in the Coleosporiaceae they become divided internally to form the metabasidial cells. In the Pucciniaceae the teleutospores may be free or united but do not form crusts, layers or columns which are characteristic of the Melampsoraceae.

Family Pucciniaceae

The teleutospores are usually pedicellate, sometimes formed in groups of three or more per pedicel (*Uromycladium*). In *Uromyces* the teleutospores are unicellular; in *Puccinia* they are two-celled; in *Phragmidium* they are phragmosporous and warted, and the pedicel is surrounded by a thick gelatinous sheath (Fig. 76).

Family Melampsoraceae

The teleutospores are united laterally into crusts or columns. *Cronartium ribicola*, responsible for blister disease of white pine, is a long-cycled heteroecious rust with aecidia giving the blistered appearance to the alternate host, white pine. The primary hosts are currant and gooseberry, on which the long columnar teleutosori and small hemispherical uredosori are formed.

Family Coleosporiaceae

The teleutospores are laterally united, forming crust-like layers. Karyogamy and meiosis occur within the teleutospore which then becomes internally four-celled. The most notable genus is *Coleosporium*, and the common species are heteroecious.

Uredinales Imperfecti

The two most notable form-genera are *Uredo*, for unconnected uredial states, and *Aecidium* for unconnected aecidial states. In *Aecidium* the peridia of the aecidia are short and straight-walled; there are several other aecidial form-genera distinguished by their types of peridia. Because Uredinales Imperfecti are clearly recognizable as states of rusts, they are classed with the Uredinales rather than with the other Fungi Imperfecti (Deuteromycotina).

Order Ustilaginales

The Ustilaginales or smut fungi are very important pathogens of crop plants and ornamentals; e.g., *Ustilago maydis* on maize, *Ustilago cynodontis* on couch lawn grass, *Tilletia caries* and *Tilletia foetida* on wheat, and *Urocystis cepulae* on onion.

These fungi are called 'smuts' because they form dry, dusty, smutty spore-masses as the most obvious stage in the life-cycle. Nearly all are parasites of flowering plants, but they are not necessarily obligate parasites; many can be grown saprobically in culture through the complete life-cycle. Many give rise to systemic infection in perennial plants. They inhabit stems, leaves and floral parts, and in one genus (*Entorrhiza*) form galls on roots. The mycelium is intercellular, sometimes forming intracellular haustoria. The teleutospores are produced in a **sorus** which may be naked but more often is covered by a membrane composed of fungal cells alone, or of fungal and host tissue. The inner mass of the sorus may be composed of teleutospores and hyphal fragments only, or may be traversed by sterile fungal threads united to form a **columella,** or by threads of host and fungal tissue.

The teleutospores are unicellular and free, or united in spore-balls which may contain both fertile and sterile cells. They are usually coloured and thick-walled, frequently ornamented, and sessile. They are formed from a dicaryotic secondary mycelium which aggregates in parts of the host and then becomes profusely septate, each cell rounding off into a teleutospore. These spores are the most important means of disseminating most smuts. The young teleutospore is dicaryotic and acts as a probasidium in which the nuclei fuse; thus the mature cell has a single diploid nucleus. On germination (Fig. 78) the teleutospore gives rise to a variable metabasidium which may sometimes be limited in growth, but often branches irregularly. Meiosis occurs in the metabasidial cells which may then branch or bud off small sessile basidiospores; these in turn may produce chains of budded conidia, sometimes resembling a pseudomycelium. The metabasidium and basidiospores are not as clearly similar to the auricularious types as in the Uredinales and are often known as a **promycelium** and **sporidia,** respectively, to point the difference. Ainsworth and Sampson (1950) use the term sporidium for all the accessory spores of smuts, i.e. those formed from metabasidia, those from hyphae, and also the bud-like conidial forms. It is quite probable that the Ustilaginales and Uredinales should not be closely allied in classification.

There are no sex organs in the Ustilaginales. Most species are heterothallic, with heterothallism of the bipolar or tetrapolar types occurring. Plasmogamy takes place by fusion of any two compatible cells, or basidiospores, or conidia, or of spores with compatible hyphae.

Classification of Ustilaginales

The way in which the teleutospores germinate (Fig. 78) is the primary feature used for delimiting families. In the Ustilaginaceae, the metabasidium becomes transversely septate, often branches, and buds off lateral and terminal basidiospores. In the Tilletiaceae, the metabasidium is aseptate (but may eventually become adventitiously septate as the cytoplasm

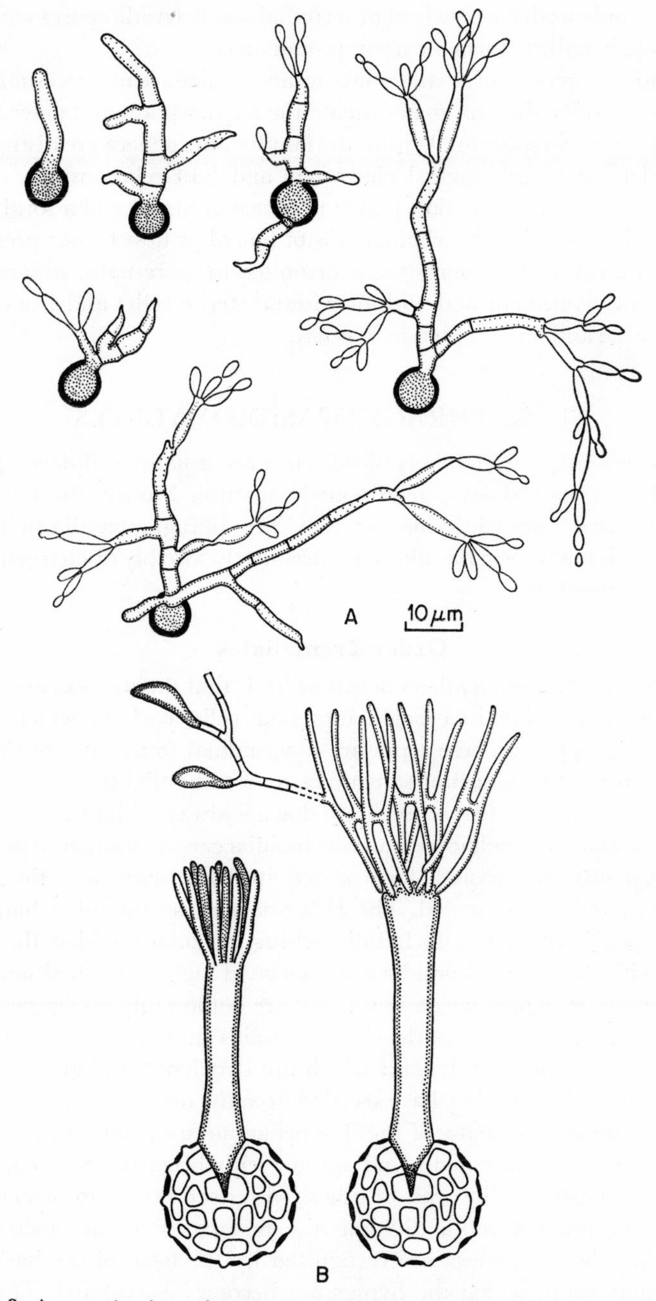

Fig. 78. A, germinating teleutospores of *Ustilago cynodontis*; B, (after Tulasne) germinating teleutospores of *Tilletia* sp. with anastomosed basidiospores giving rise to mycelium bearing conidia.

regresses) and produces a whorl of terminal sessile basidiospores which may anastomose in pairs to form dicaryotic common cells.

Keys to the genera of Ustilaginaceae and Tilletiaceae are usually combined, as it is often difficult to germinate the teleutospores and determine the nature of the metabasidium. Within the families, the genera are distinguished on a variety of morphological characters and host relationships, such as: position of the sorus in the host plant; presence or absence of a soral sheath; nature of the soral sheath, whether it is of fungal or host tissue; presence or absence of a columella; spores single or united in spore-balls, which may be persistent or evanescent and with or without sterile cells; and size of spores (usually large and ornamented in *Tilletia*).

CLASS PHRAGMOBASIDIOMYCETES

The basidiocarps in members of this class are usually well developed and frequently, but not always, gelatinous in texture. Mostly, the basidia are formed in well defined hymenia; they are phragmobasidia with large variable sterigmata. The basidiospores are mostly forcibly discharged and are frequently repetitive.

Order Tremellales

The chief feature of members of this order is that the metabasidia become cruciately divided by three septa into four cells, each of which bears a sterigma (Fig. 74, J). These septa are longitudinal to the axis of the metabasidium in most genera, but sometimes tend to be slightly oblique; in the genus *Patouillardina* the first septum laid down is always oblique.

In the families Tremellaceae and Sirobasidiaceae the basidiocarps develop gymnocarpously, the basidia form a definite hymenium, and the basidiospores are forcibly discharged. The Hyaloriaceae, on the other hand, have angiocarpous fruitbodies with basidia arising irregularly within the fructifications; with the enclosed basidia are associated highly modified sterigmata, and basidiospores which lack a hilum and are not forcibly discharged. In the Sirobasidiaceae, the basidia develop in chains and bear relatively large, sessile, spore-like protosterigmata which are deciduous and give rise to the basidiospores only after they have seceded from the metabasidia.

The basidiocarps in most of the Tremellaceae are gelatinous; many are whitish, but it is common to find species with brightly coloured, yellow, orange or reddish basidiocarps, or others with deep brown to blackish ones. In a few genera the septa dividing the metabasidium are incompletely formed. Mostly, the genera are recognized by the form of the basidiocarp and the manner in which the hymenium becomes convoluted. Thus there are genera with effused, clavarioid, pileate or cupulate basidiocarps, and the hymenium may be wrinkled, smooth, hydnoid or even poroid. Probably the commonest genera of the Tremellaceae are *Tremella* with an effused, cushion-shaped or foliose basidiocarp and spherical spores, and *Exidia* which differs mainly in having cylindrical or elongated spores.

Order Auriculariales

In members of this order, the probasidium puts out, or enlarges directly into, a cylindrical metabasidium which becomes horizontally septate and one–four celled, each cell bearing a sterigma and a basidiospore. The sterigmata are typically stout and variable (Fig. 74, G, I).

In the Phleogenaceae the basidiocarps are stalked and pileate, usually rather small, hemiangiocarpous, and either dry or gelatinous in texture. The sterigmata tend to be reduced in size, a feature correlated with the fact that the basidia are enclosed in the developing fructification. The Auriculariaceae are typically gymnocarpous and seldom pileate; even when pileate their basidiocarps do not have a distinct stipe but instead are narrowed at the point of attachment. Some of their fruitbodies are gelatinous, others more arid. A number of genera have parasitic members, but the majority are saprobes. The form of the fructification may vary from effused, effuso-reflexed, cushion-shaped, lobate to fan-like or cupulate. In some instances the probasidium is more or less spherical and the metabasidium develops as a cylindrical extension from it; the probasidium may then disappear or persist as a virtually empty sac at the base of the metabasidium. In other instances the probasidium is clavate and, although it generally elongates, its place is taken by the metabasidium in the mature basidium. All these features are used taxonomically to delimit genera.

Order Septobasidiales

In this order the basidiocarps are not gelatinous and they are always formed in association with scale insects. The metabasidia are like those of the Auriculariales, with a probasidium which often persists as a sac at the base of the mature basidium (Fig. 74, H). Two genera are recognized: *Septobasidium* and *Uredinella.* Their extraordinarily complex associations with scale insects make a fascinating story which was admirably investigated by Couch (1938) in one of the best mycological monographs ever written; it has been retold by Christiansen (1961) in a most readable form prefaced with the remark that 'while what follows may be as implausible as a fairy story, it has the most excellent evidence behind it'.

CLASS HOLOBASIDIOMYCETES

The basidiocarps in members of this class are mostly conspicuous and sometimes very large indeed. While a few members are gelatinous or waxy in texture, most are typically fleshy to tough and woody or corky. The basidia are holobasidiate, occasionally with adventitious septa. Various forms of sterigmata are encountered but in most orders they tend to be relatively small, not voluminous or highly variable. The basidiospores may or may not be forcibly discharged, depending upon whether the basidia are enclosed at maturity. Repetition of the basidiospores is known only in a few genera.

Order Dacrymycetales

The basidia in members of this order are always reminiscent of a tuning-fork, having a cylindric-clavate metabasidium and a pair of stout variable sterigmata (Fig. 74, D). All members are saprobes and most are gelatinous and brightly coloured. In the genus *Dacrymyces* the basidiocarps are discoid or cushion-shaped, with a smooth or wrinkled hymenium on the outer side; the basidiospores may become septate and bud off conidia at maturity. In *Calocera* the basidiocarps are club-shaped to coralloid while in *Dacryopinax* they are spathulate. In *Guepiniopsis* the basidiocarps are spathulate at first but become discoid to cupular at the apex; the hymenium is unilateral on the superior surface of the fertile apex, and there are sterile bottle-shaped cortical cells on its inferior surface. All these forms occur mainly on dead wood or on cones.

Order Tulasnellales

Members of the Tulasnellales may be saprobic or parasitic, and form effused waxy to web-like fructifications on rotten wood, soil, or living or dead plant parts. Several species, with mycelial states assigned to the form-genus *Rhizoctonia*, form mycorrhizal associations with terrestrial orchids. As conceived here, this order comprises holobasidial fungi with broad or subspherical basidia, stout or inflated variable sterigmata and repetitive basidiospores. Two families are recognized: the Tulasnellaceae (example *Tulasnella*) in which the sterigmata are strongly inflated and are cut off by basal adventitious septa often to become deciduous, and the Ceratobasidiaceae in which the sterigmata are broad and finger-like but not inflated or deciduous (Figs. 74, E, F). Examples of the Ceratobasidiaceae are *Ceratobasidium* in which the basidia are subspherical and borne on narrow hyphal pedicels, and *Thanatephorus* in which the basidia are cylindrical or clavate and borne on pedicels nearly the same width as the metabasidia. *Thanatephorus cucumeris*, the perfect state of *Rhizoctonia solani*, is very widely distributed in soil and is pathogenic to an extremely wide range of host plants. There is recent evidence that species of *Ceratobasidium* may parallel *Thanatephorus* in this respect. The taxonomic position and limits of these families are highly controversial, depending as they do upon the definition given to a holobasidium or a phragmobasidium and to the weight given to this feature in classification. Most authors regard these families as not very closely related members of the Tremellales (Heterobasidiomycetes).

Order Brachybasidiales

Brachybasidium pinangae, the best known species representing this order, is a parasite of palm leaves. Its basidia erupt from stomata on the lower surfaces of the leaves and form small tuft-like sori composed of probasidia with thickened walls. Further development takes place with the emergence from the probasidium of a cylindrical metabasidium bearing two sterigmata. A marked constriction separates the metabasidium from the persistent probasidium. Such basidia, in which development takes place in two stages, morphologically delimited by a constriction, are said to be **dimerous;** basidia with this form of development are to be found in several other groups of Basidiomycotina.

Order Exobasidiales

Members of the genus *Exobasidium* are parasitic on Ericaceae, *Azalea*, *Rhododendron* and several other plants; *E. vexans* is especially important as the cause of blister blight of tea. The intercellular mycelium causes hypertrophy of the host tissues. There is no true basidiocarp, the basidia being formed from somatic hyphae beneath the host epidermis and bursting through it to produce an effused layer of indefinite extent. The effect is very similar to that caused by *Taphrina*, a Hemiascomycete, and indeed the two genera are held to be closely allied in a recent classification.

Order Aphyllophorales

Where modifications of the Friesian classification recognize about half a dozen families of Aphyllophorales based mainly on the shape of the fruitbody and the convolution of the **hymenophore** (the part of the pileus or fructification bearing the hymenium), a recent authoritative work (Donk, 1964) divides this order into twenty-one families. This great change has come about largely because the problem of homoplasy is now more clearly appreciated and because both microscopic and macroscopic characters, rather than predominantly macroscopic ones, are used to define the families. A family of Aphyllophorales is now defined by means of a large number of external and internal features in combination, rather than by depending upon one or two striking superficial features to set it apart from other families. What formerly appeared to be exclusive characters now diminish in importance and become common to several families. Short keys become difficult to construct and fail to reveal the full range of variability within the families; classification becomes much harder to grasp than previously. The same remarks apply in the Agaricales (dealt with below). These two orders comprise most of the larger and conspicuous Basidiomycotina. Desirable though it may be to classify them on field characters which can be grasped by the beginner in mycology, it would nevertheless be unrealistic to do so in the light of modern progress — one would simply be returning to the Friesian system already explained above. To illustrate Donk's classification of the Aphyllophorales I thought it worthwhile to include the following key even though it will certainly be found imperfect and difficult to use in practice.

Families of Aphyllophorales

1. Fruitbody pileate, dimidiate or stipitate; hymenophore poroid; trimitic; with clamp connexions; spores coloured, with an inner brown layer ornamented with spines which pierce the hyaline outer layer, often apically truncated *Ganodermataceae*

   Not combining these characters 2.

2. Fruitbody pileate, laterally stipitate or subsessile, fleshy; hymenophore of densely crowded tubes, free from one another; monomitic; spores hyaline, smooth *Fistulinaceae*

   Not combining these characters 3.

3. Fruitbody cupulate in origin, attached by a narrow stalk-like base, sometimes becoming discoid, sometimes (as in *Schizophyllum*) with adjacent cups proliferating and uniting at the margins to form a pseudolamellate agaricoid fruitbody with radially split 'gills'; monomitic; spores hyaline, non-amyloid (including some 'Cyphellaceae') *Schizophyllaceae*

   Not combining these characters 4.

4. Fruitbody effused or effuso-reflexed; hymenophore wrinkled to tuberculate, with fertile cushions separated by narrow sterile fissures filled with amorphous material; catahymenium with dendrohyphidia whose apices are brown and knob-like; monomitic; with clamp connexions; spores smooth, hyaline, non-amyloid *Punctulariaceae*

   Not combining these characters 5.

5. Spores coloured and cyanophilous; spore-wall two-layered
   (*a*) Spores smooth, with cyanophilous inner layer; fruitbody not greening with 10% $FeSO_4$ *Coniophoraceae*
   (*b*) Spores usually roughened, rarely smooth, with cyanophilous outer layer; fruitbody turns green with 10% $FeSO_4$ *Gomphaceae*

   Spores coloured or hyaline, but not cyanophilous 6.

6. (*a*) Context xanthochroic, i.e. yellow-brown or darker and permanently colouring dark in 10% KOH; clamps always absent; hyphae usually coloured; setae present in most species; asterosetae or dichohyphidia present in some; spores usually smooth, if ornamented then even in outline *Hymenochaetaceae*
   (*b*) Context pallid to usually dark, darkening or becoming greenish in KOH; clamps present or absent; spores usually ornamented and uneven in outline *Thelephoraceae*

   Not combining the characters of either 7.

7. Spores amyloid
   (*a*) Spores thin-walled
   (i) Fruitbody usually large; hymenophore poroid or clavarioid; spores with conspicuous strongly amyloid spines and crests *Bondarzewiaceae*
   (ii) Fruitbody with dentate or lamellate (lacerate-dentate gills) hymenophore; spores smooth or minutely sculptured; gloeocystidia darkening in

sulpho-aldehyde; generative hyphae thin-walled, with clamps *Auriscalpiaceae*

(*b*) Spores thick-walled

(i) Fruitbody pileate or clavarioid, with dentate or smooth hymenophore; spores smooth or minutely asperulate; gloeocystidia not darkening in sulpho-aldehyde; generative hyphae thin-walled to thick-walled, with clamps *Hericiaceae*

(ii) Hymenophore dentate, corky or woody in texture; spores smooth; generative hyphae thin-walled to thick-walled and clamped *Echinodontiaceae*

(*c*) If without the above combinations of characters, see under *Stereaceae*, *Corticiaceae* and *Polyporaceae* 12.

Spores non-amyloid 8.

8. Hymenophore dentate; fruitbody stipitate-pileate

(*a*) Generative hyphae thin-walled, with clamps; spores smooth *Hydnaceae*

(*b*) Generative hyphae thin- or thick-walled, without clamps; spores echinulate *Bankeraceae*

Hymenophore not dentate 9.

9. Fruitbody funnel-shaped (infundibuliform) or tubular; generative hyphae thin-walled, inflating; hymenophore smooth, wrinkled or folded; spores smooth, hyaline *Cantharellaceae*

Not combining these characters 10.

10. Basidia regularly two-spored; sterigmata generally stout, strongly curved; fruitbody clavarioid; generative hyphae inflating *Clavulinaceae*

Not combining these characters 11.

11. Fruitbody clavarioid, erect, simple or branched; hymenium amphigenous; hymenophore smooth or becoming wrinkled *Clavariaceae*

Fruitbody erect, branching into flattened or wavy lobes; hymenium inferior on horizontal surfaces of lobes or amphigenous on erect surfaces; hymenophore smooth; monomitic; with inflating hyphae and vascular hyphae *Sparassidaceae*

Not combining the above characters 12.

12. Fruitbody strictly effused; frequently monomitic; hymenophore smooth, wrinkled, tuberculate or dentate *Corticiaceae*

Fruitbody effused, reflexed or stipitate-pileate; monomitic, dimitic or trimitic; hymenophore poroid, irpicoid or lamellate *Polyporaceae*

Fruitbody effuso-reflexed to stipitate-pileate; dimitic or exceptionally trimitic; hymenophore smooth or ribbed (costate); typically with context differentiated into a trichoderm, a horizontal closely woven cortex, and a looser ascending layer of hyphae terminating in the hymenium *Stereaceae*

Some of the effused Aphyllophorales, whose fruitbodies may merely resemble a splash of dingy paint on wood or bark, are not exciting to observe with the naked eye, but even the most unprepossessing may contain a wonderful assortment of sterile hymenial structures (Fig. 73) — various sorts of cystidia, hyphidia and setae — which make their microscopic study fascinating and rewarding. Little is known about the functions of these sterile elements.

The Aphyllophorales comprise a vast number of genera (whose limits are not always agreed upon) and species, most of which are saprobes. Relatively few, except those which cause rot of standing trees and decay of worked timber, are of direct economic importance.

Two general types of decay of timber are known: the **'white-rots'** and the **'brown-rots',** which are distinguished by the chemical effects produced upon the timber by the fungi concerned. The fungi responsible for 'white-rots' produce both oxidizing and hydrolyzing enzymes which attack all elements of the wood, including lignin. The fungi causing 'brown-rots' produce hydrolyzing enzymes only, attacking the cellulose and pentosans but leaving the lignin almost intact. Thus the type of rot may be characterized chemically, by testing for the presence or absence of extracellular oxidases in cultures of the fungus. 'White-rot' fungi have these oxidases, which can be detected in cultures containing gallic acid or tannic acid by the presence of dark diffusion zones surrounding the fungus; 'brown-rot' fungi do not have the oxidases and the medium remains clear. Alcoholic gum guiacum solution can be applied directly to the culture to test for extracellular oxidases; rapid blueing of the reagent indicates the presence of oxidases and a 'white-rot' fungus. Such tests are useful in the identification of species in culture and have also been used (Nobles, 1958) in a phylogenetic and taxonomic treatment of the Polyporaceae. Nobles regards the absence of extracellular oxidase combined with bipolar heterothallism as indicating more primitive species. The presence of oxidases, tetrapolar heterothallism, and regular clamp connexions were taken to indicate a higher phylogenetic level. On this basis, and with a consideration of hyphal and basidiospore characters, Nobles re-grouped species of Polyporaceae into thirty-six groups which may represent more natural taxa than some of the present genera.

Many species of Polyporaceae cause heart-rot or butt-rot of standing trees, resulting in considerable economic loss. One such fungus is *Fomes annosus*, whose spores germinate readily on freshly-cut stumps of conifers, establishing a mycelium which spreads by means of mycelial strands to the base of neighbouring living trees, there producing a serious butt-rot. A most elegant biological control of this fungus has been demonstrated by Rishbeth (1963), who sprayed a suspension of spores of *Peniophora gigantea* (Corticiaceae) on conifer stumps to compete with those of *Fomes annosus*. *P. gigantea* grows fast enough to suppress the growth of *F. annosus*, but does not spread through the soil to living trees.

Various fruit trees may be severely damaged by *Chondrostereum purpureum* (Stereaceae) whose attack on stems and branches is revealed by the appearance of 'silverleaf' symptoms, caused by the production of airspaces beneath the epidermis of the leaves. *Corticium salmonicolor* (Corticiaceae) is responsible for pink-rot of rubber and other trees. An intriguing relationship, not yet fully worked out, exists between some species of *Amylostereum* (Stereaceae), *Sirex* woodwasps and conifers. The woodwasps carry arthrospores of *Amylo-*

stereum in mycangia within their bodies which communicate with the ovipositors; thus when eggs are deposited beneath the bark of conifers some fungal inoculum is transferred as well. The fungus grows in the wood near the site of oviposition and by softening it, probably assists the larvae to tunnel. Despite the fact that the fungus spreads slowly, and apparently not far, symptoms of needle-browning towards the top of the tree are observable very soon after oviposition. This suggests that the symptoms are caused by chemicals taken up in the transpiration stream, but the details are not yet known.

In nursery beds, *Thelephora terrestris* (Thelephoraceae) often grows saprobically in soil but may encircle pine seedlings with its pilei and kill them.

Two members of the Coniophoraceae, *Coniophora puteana* and *Serpula lacrymans*, are extremely important in causing most of the devastating 'dry-rot' of worked timber in buildings. Dry-rot is a misnomer, since timber that has been seasoned to a moisture content of below about 20% of its dry weight and then kept ventilated is not susceptible to attack by decay fungi. The timber must be wet or moist before either of these fungi can attack it; but after the fungus has become established it can produce its own water requirements by chemical breakdown of fairly dry wood, which finally becomes dry and crumbly. Both these fungi are able to travel far in buildings, reaching new areas of woodwork by means of their rhizomorphic strands which can traverse soil or even penetrate through mortar. Both are extremely destructive and can usually be eradicated only by replacing decayed timber and treating the whole area with fungicides before installing new, treated timber.

In the warm and humid atmosphere of deep mines a variety of fungi may decay the mine-props. *Poria vaillantii* and *Poria incrassata* (Polyporaceae) are perhaps the commonest of these, and may destroy props within a year of their having been placed in the mines. Another not uncommon species is *Ganoderma lucidum* (Ganodermataceae); in the open, the fruitbodies of this species are usually stipitate-pileate with the surface of the stipe and pileus covered with a rich brown natural lacquer. In mines, however, the absence of light induces the formation of strange antler-like fruitbodies, with small fertile patches confined to the extreme tips of the branches instead of occupying the whole under surface of the pileus.

Most of the Aphyllophorales have a tough context which renders them inedible, but a few are sufficiently fleshy to be eaten. Among these is *Sparassis crispa* (Sparassidaceae) which looks not unlike a cauliflower head at the base of a tree; *Fistulina hepatica* (Fistulinaceae), the 'beefsteak fungus', so-called because of its reddish colour and edibility; and *Cantharellus cibarius* (Cantharellaceae), the 'chantarelle' of European mycophagists.

Order Agaricales

In most Agaricales the basidiocarp is fleshy and umbrella-like, with a central stipe and a pileus; some, however, are sessile and laterally attached. Lamellae (gills) or tubes bearing the hymenium are present on the lower surface of the pileus (Fig. 79).

Most of the forms bearing a tubular hymenophore fall in the family Boletaceae (Fig. 72, K), which can be distinguished from the poroid members of the Aphyllophoraceae by the fact that the tubes are typically fleshy and easily separable from the pileus. Some of the Boletaceae form mycorrhizas with trees and may be important in the establishment of pine plantations. Even though they are often referred to as 'toadstools' most boleti are edible, though many people do not find them palatable because they tend to become slimy when cooked. A few species, such as *Boletus satanus*, are poisonous. The pilei of some species of boleti in Madagascar and East Africa, such as *Phlebopus colossus* which is suspected of being poisonous, and of *Boletus portentosus* in South Australia, may be as much as 43 cm in diameter and weigh several kilograms when fresh.

The lamellate members of the Agaricales are commonly called 'mushrooms' or 'toadstools' according to whether they are regarded as edible or not; mycologists prefer to call all of them 'mushrooms' or 'agarics' since, apart from the hard school of experience, there is no simple test for edibility other than to be able to identify the species (not necessarily naming it) and to know its reputation. Relatively few agarics are poisonous, and most fatalities can be attributed to species of the genus *Amanita*. Some normally harmless agarics can cause symptoms resembling those of nitrite poisoning when they are eaten at a meal with alcohol; the principal offenders in this regard are the black-spored *Coprinus comatus* and *Coprinus atramentarius*.

The most popular edible mushrooms in western countries are *Agaricus bisporus* (the cultivated mushroom), *A. campestris* (the field mushroom; Fig. 79, A) and *A. arvensis* (the horse mushroom). One common species of *Agaricus*, poisonous to some susceptible people, should be noted; this is *A. xanthodermus*, much like the well known field mushroom but easily recognized by the fact that any part of the fresh stipe or pileus immediately turns saffron yellow when rubbed or bruised. In south-east Asia and neighbouring islands, *Volvariella esculenta* is grown in large quantities on rice straw. In Japan the commercial mushroom is the 'shii-take', *Tricholoma edodes*, which is grown on oak or hornbeam logs. These are soaked in water and then softened by pounding before the plugs of mycelium used as inoculum are placed into holes bored close together in the wood. The logs are watered periodically and kept in the shade of forests or in greenhouses until the crop develops, usually after about 2 years.

In the genus *Amanita* (Fig. 79, B, C, D, E) the young fruitbodies are

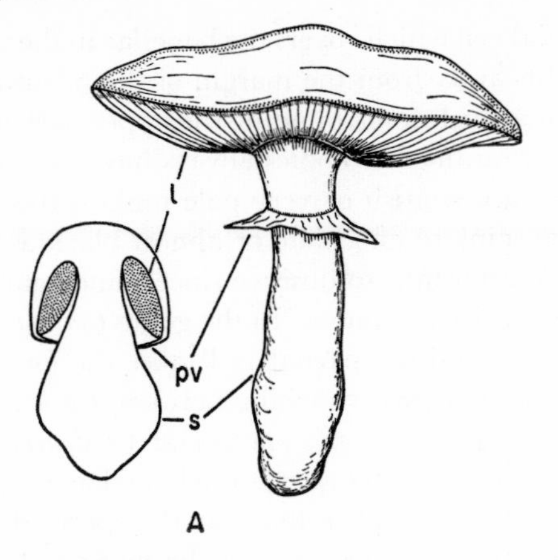

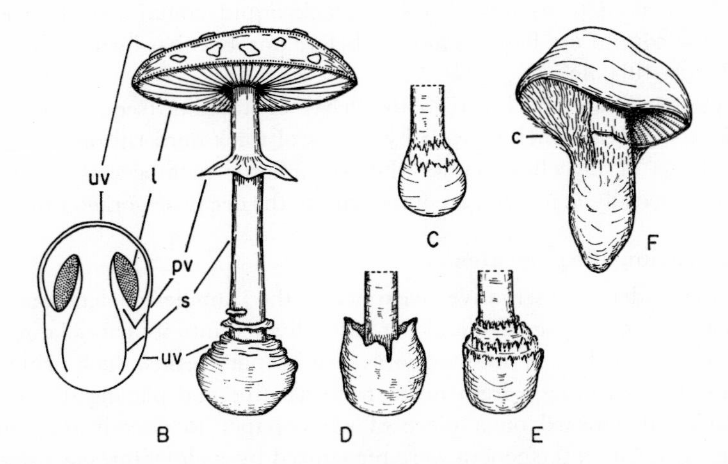

Fig. 79. A, *Agaricus campestris*, young and expanded fruitbodies; B, *Amanita pantherina*, young and expanded fruitbodies; C, volva of *Amanita rubescens*; D, volva of *Amanita phalloides*; E, volva of *Amanita muscaria*; F, *Cortinarius* sp. showing partly expanded pileus with the cortina; *c*, cortina; *l*, lamellae; *pv*, partial veil; *s*, stipe; *uv*, universal veil forming scurf or warts on pileus and the volva at the base of the stipe.

covered by a **universal veil** which splits as they expand. Remnants of the veil often remain as scurf or warts dotted about the pileus and always as a sheath or **volva** surrounding the base of the stipe. The **volva** has the appearance of a membranous cup in *Amanita phalloides*, a series of membranous partial rings in *A. pantherina* and a series of dentate projections in *A. muscaria*; in some species it may even be reduced to fine scales. In *A. phalloides*, the most poisonous of all species, the pileus is smooth to striate, and of a light greenish hue. All species of *Amanita* have white to cream-coloured lamellae and all have a partial veil which covers the lamellae in the young unexpanded fruitbodies but splits away from the margin of the pileus as it expands, and remains as a ring or **annulus** surrounding the upper part of the stipe.

In the genus *Agaricus* the fruitbodies always have an annulus but never a volva; the lamellae are whitish or very pale pink in the young fruitbodies, but change through pink to chocolate or almost black as they mature. The change in colour is important to observe since some agarics whose gills are pink and remain pink, are poisonous. In the genus *Cortinarius* (Fig. 79, F) the lamellae are brownish and the partial veil takes the form of a cobweb-like series of threads (the **cortina**) stretching between the stipe and the margin of the pileus in young specimens; species of *Cortinarius* are regarded as poisonous or at least suspect. In *Coprinus* the lamellae are pale pink or white at first but rapidly darken to pure black as the spores develop, and at the same time they become autodigested from the margin of the pileus inwards so that the flesh turns into drops of black liquid containing the spores. Naturally, coprini are best to eat only before autolysis has begun and while the gills are still pale.

Some agarics are concerned in the decay of standing trees, for example *Armillariella mellea*, which spreads by means of thick dark rhizomorphs and forms sheets of mycelium under the bark. Some members of the genus *Lentinus*, especially *L. lepideus*, are important in the decay of worked timber.

Classification of Agaricales

In the older, conservative treatments the families Boletaceae and Agaricaceae were recognized and the latter divided into six tribes largely on the basis of the colour of massed spores seen in spore-prints. Such prints can be made by detaching a fresh pileus from its stipe and placing it with the gills facing downward on a piece of white paper to deposit the spores. Within these tribes, the genera were recognized by such features as presence or absence of an annulus and/or volva, mode of attachment of the gills to the stipe, presence of a milky juice in the flesh, insertion of the stipe (whether central, excentric or absent), and texture of the stipe and pileus. Even with a relatively simple artificial system such as this, agarics have always been difficult to determine without close study and without a field knowledge of species gained largely as a disciple of an experienced agaricologist. The necessity for close and painstaking study is greatly accentuated when one attempts to comprehend and make use of the modern classification of Agaricales proposed by Singer (1962*a*, 1962*b*) in his great work on 'The Agaricales in Modern Taxonomy' and 'Keys for the Determination of the Agaricales'.

The reader is referred to these books for details of the classification and of the sixteen families of Agaricales recognized by Singer. The main point I wish to make here is that once again the older classifications relying on a few macroscopic features have proved inadequate, and natural groupings have to be sought on the basis of combinations of microscopic and macroscopic characters, with the former probably more definitive. Among the microscopic characters used by Singer for this purpose are: spore ornamentation and amyloidity; distribution of clamp connexions; structure of the epicutis, the outermost layer of the pileus surface; cystidial types; and especially the tissue arrangements to be found in the context (or **trama**) of the lamellae.

In a trama of **homoiomerous** structure the fundamental tissue is composed of hyphae only; this applies in the majority of families. In the Russulaceae, however, the trama is **heteromerous,** i.e. composed of hyphae and either large rounded cells known as **sphaerocysts** (in the genus *Russula*; Fig. 80, A), or laticiferous hyphae (in *Lactarius*). Homoiomerous

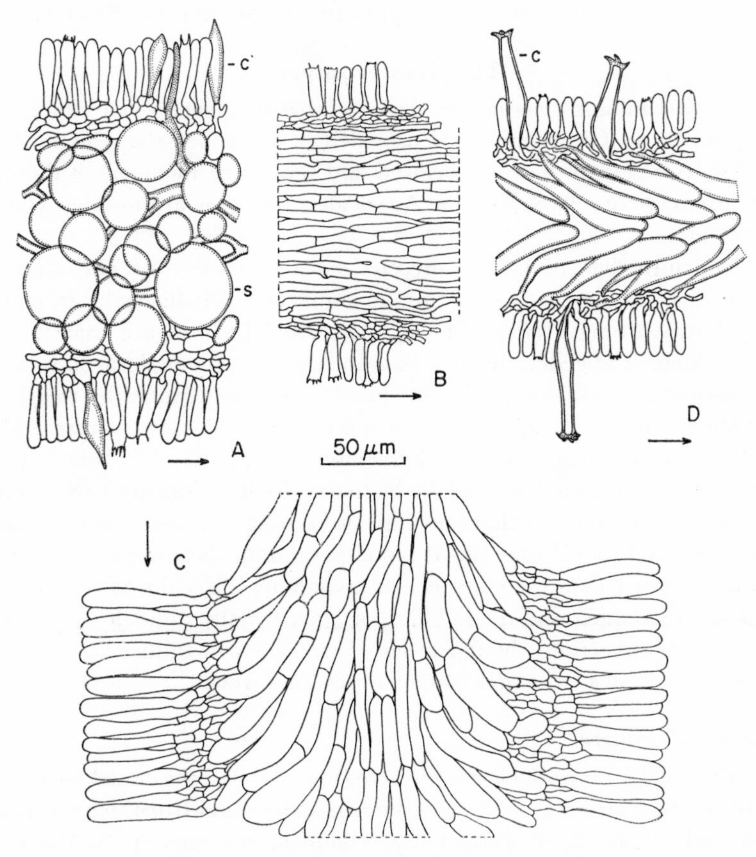

Fig. 80. Gill structure in Agaricales. A, heteromerous trama of *Russula* sp., showing hyphae and sphaerocysts (*s*), basidia and cystidia (*c*); B, C, D, homoiomerous trama; B, regular trama in *Rozites australiensis*; C, bilateral trama in *Amanita* sp.; D, inverse trama in *Pluteus cervinus*, with cystidia (*c*). The direction of the gill edge is shown by an arrow in each case.

groups can be further subdivided by considering the directional arrangement of the hyphae in the lamellar trama (Fig. 80). The trama may be **irregular,** without a marked parallel or divergent pattern to the direction of the hyphae, but instead with either irregularly interwoven hyphae or with a mixture of hyphae and sphaerocysts. In a **regular** trama, the hyphae are more or less parallel. In instances where the hyphae diverge two types of trama are recognized: bilateral and inverse. The **bilateral** trama has a narrow central strand of parallel hyphae (the **mediostratum**) from which parallel branches diverge and curve towards the subhymenium, joining it nearer to the gill edge than the point of departure. In an **inverse** trama, characteristic of the Amanitaceae, the outer layer of hyphae diverging from the mediostratum joins the subhymenium further away from the gill edge than their point of departure. These terms referring to gill structure are illustrated in Fig. 80.

The 'Gasteromycetes'

The remaining orders of the class Holobasidiomycetes can conveniently be grouped as 'Gasteromycetes' provided it is realized that this no longer represents a class of Basidiomycotina and that some members of the group are more closely related to some Agaricales than they are to other Gasteromycetes.

The fruitbodies of Gasteromycetes have a variety of curious forms, including those known by the common names of puffball, earthstar, phalloid and birdsnest-fungus (Figs. 81–83). The chief features are that they are holobasidiate and generally have basidiocarps with an angiocarpous form of development. These either remain closed until disintegrated by weathering or animals, or they expose the spores only after they are mature and the basidia have disintegrated. The hymenium is seen only in young stages of the fruitbody. Although several distinctive types of basidia are known among Gasteromycetes, the fact that they have to be sought in immature specimens, as well as the ridicule that Lloyd (1902)poured on their use as a taxonomic character, has led to their neglect in taxonomy. They deserve to be systematically examined and used. The spores are usually difficult to germinate but a small proportion, about 0·1–0·2%, can slowly be induced to germinate in the presence of a yeast, *Rhodotorula* sp. (Bulmer, 1964).

General structure

In the puffball-type of Gasteromycete (Figs. 81, 82), the closed basidiocarp is bounded by an outer wall tissue known as the **peridium,** which may be composed of one–three distinct layers known, respectively, as the **endoperidium** (on the inside), the **mesoperidium,** and the **exoperidium** (on the outside). It may have a natural opening (**stoma**) or may weather away to release the spores. The peridium encloses the fertile part called the **gleba.** In some types the gleba is divided, at least at first, by plates of sterile tissue

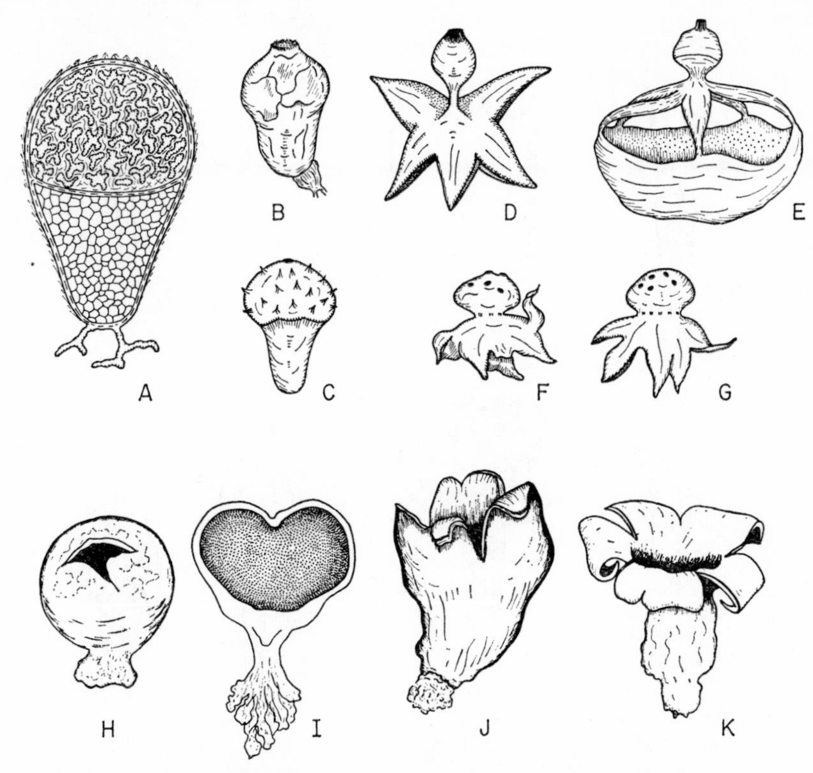

Fig. 81. Puffball type of 'Gasteromycete' fruitbodies. A, (diagrammatic) section of *Lycoperdon* sp. showing fertile gleba above and sterile cellular base below the diaphragm; B, *Bovista*; C, *Lycoperdon*; D, E, *Geastrum*; F, G, *Myriostoma*; H, I, J, K, *Scleroderma* fruitbodies in various stages of splitting open (I in section).

(**tramal plates**) which form chambers; or the spores may be mixed with glebal hyphae (**capillitium**) left after the basidia have disintegrated. In some genera, a sterile **columella** traverses the gleba; or a **diaphragm** may separate the gleba from a cellular sterile base.

In the Phallales (Fig. 83) the young fruitbody is an 'egg' formed below the surface of the soil; part of the ruptured peridium is left behind in the soil as a **volva** encircling the base of a columnar or netlike (clathrate) **receptacle** which expands above soil level and bears the gleba in an exposed position.

In the Nidulariales, or 'birdsnest-fungi' (Figs. 84, 85) the peridium is globose or funnel-shaped, and segments of the gleba become rounded off into independent **peridiola.**

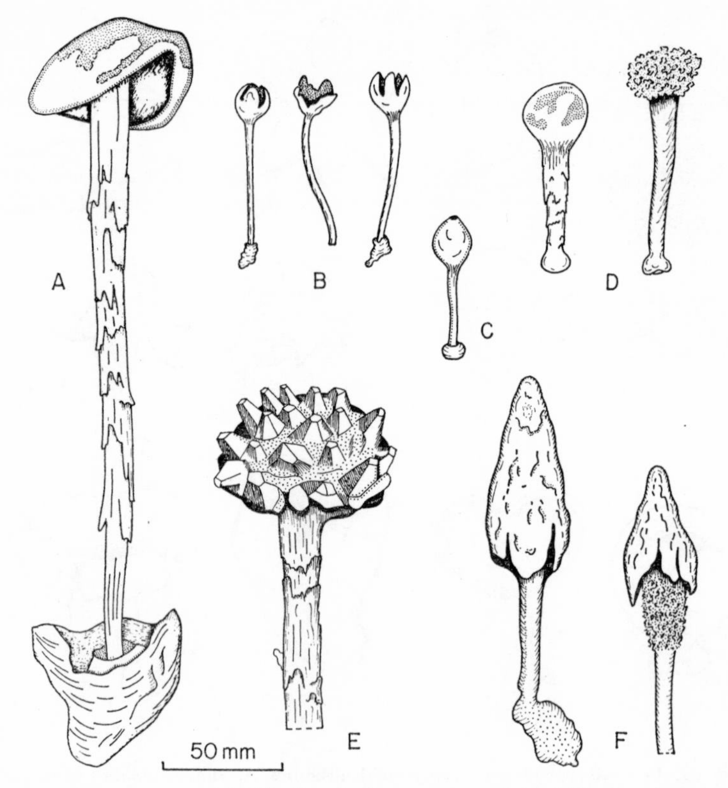

Fig. 82. 'Gasteromycete' fruitbodies. A, *Battarraea stevenii*; B, *Schizostoma laceratum*; C, *Tulostoma* sp.; D, *Chlamydopus meyenianus*; E, *Phellorina strobilina*; F, *Podaxis carcinomalis*.

Order Hymenogastrales

In this order there are two families: the Hymenogastraceae whose members have sessile fruitbodies, and the Secotiaceae with stipitate fruitbodies. The peridium has one–three layers and is indehiscent. The gleba is composed of anastomosed tramal plates enclosing cavities lined by the hymenium. There is no capillitium. Some members have a dendroid columella; *Secotium* has a simple columella extending through the gleba from the apex of the stipe. Many members of the Hymenogastrales are hypogean and superficially resemble truffles.

Order Phallales

The peridium of one–two layers at first encloses the gleba and receptacle in the 'egg' stage, but finally ruptures and remains as a volva at the base of the receptacle. The receptacle has a spongy texture and is columnar or clathrate, bearing a foetid mucilaginous spore-mass (gleba) on some part of its surface. The basidiospores are usually very small, and are distributed by insects attracted by the odour, or by splashes of water.

In the family Phallaceae (Fig. 83, A–E) the receptacle is unbranched, hollow and spongy, bearing the gleba over its apical part (*Mutinus*), or on an apical pileus (*Phallus*); in *Dictyophora*, whose species are sometimes called 'crinoline fungi', there is a lacy network (**indusium**) suspended from the margin of the pileus.

In the Clathraceae (Fig. 83, F–I) the receptacle is composed of a number of free, connivent or united branches and the gleba, with spores embedded

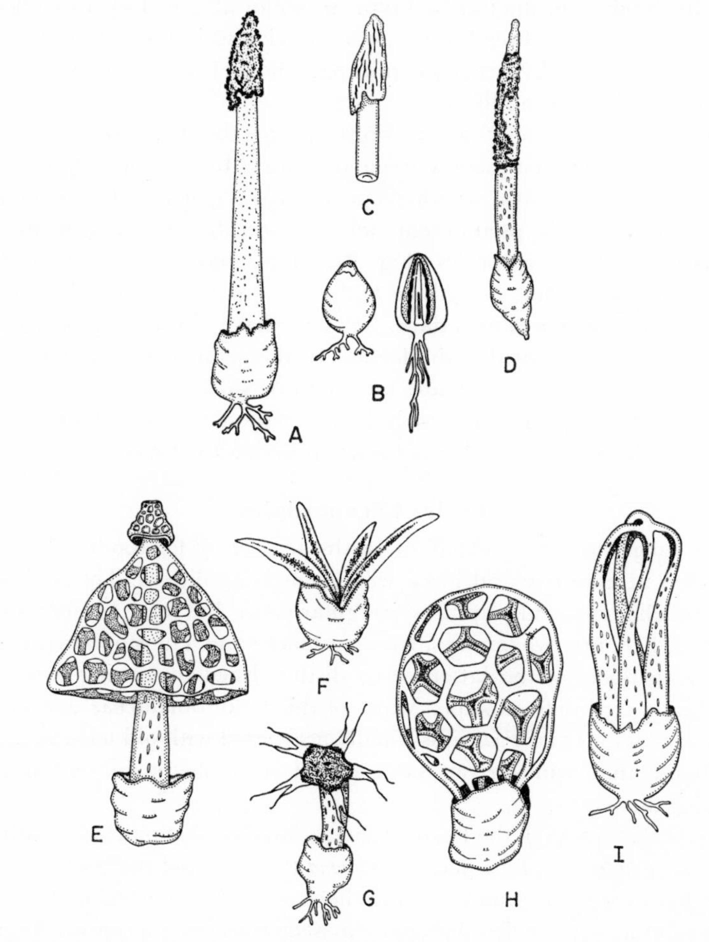

Fig. 83. Phallaceae: A, B, C, *Phallus rubicundus*: A, receptacle arising from volva and bearing the gleba on an apical pileus; B, 'egg' stage, whole and in section; C, pileus with gleba removed; D, *Mutinus caninus*; E, *Dictyophora* sp. Clathraceae: F *Lysurus* sp.; G *Aseroë rubra*; H, *Clathrus* sp.; I, *Linderiella* sp.

in mucilage, is borne on some part of the branches. In *Linderiella* the receptacle consists of simple arms united at the apex but free towards the base. In *Lysurus* the arms are connivent at the base but free at the apex; and in *Aseroë* the apex of the receptacle is discoid and horizontally expanded into a number of simple or branched arms. In *Clathrus* the entire receptacle is composed of anastomosed branches forming a hollow spherical network. The receptacle of many phalloids is brightly coloured, often pink or orange or scarlet, and the gleba is a dull olive green to blackish slimy mass.

Order Sclerodermatales

The fruitbody is of the puffball type, sessile or attached by a pseudostem. The peridium is one–three layered, often thick and indurated, and dehisces either by a stoma or by irregular splitting. The gleba is dry and powdery at maturity, and lacks a capillitium.

In *Calostoma*, the only genus representing the Calostomataceae, the peridium is carried on a pseudostem and dehisces by an apical stoma. In the Sclerodermataceae, with several representative genera, the peridium is sessile or on a short pseudostem and dehisces by irregular fissuring or weathering. *Scleroderma* species (Fig. 81, H–K) have a powdery gleba at maturity; some of these are edible and occasionally used as substitutes for truffles before the spores turn dark, but some have also been reported to be poisonous when consumed with alcohol. In the genus *Pisolithus*, the gleba is divided by persistent tramal plates into chambers filled with powdery spores. *Pisolithus tinctorius* forms a mycorrhizal association with eucalypts; extracts of its fruitbodies were formerly used as a yellow dyestuff for textiles.

Order Lycoperdales

These fungi have the puffball or earthstar types of fruitbodies. The gleba is powdery at maturity and has a well formed capillitium. The peridium is either indehiscent and ruptures irregularly, or has one or more apical stomata. There are two families, Lycoperdaceae and Tulostomataceae (Figs. 81, 82), members of the latter being distinguished by having a peridium carried on a well defined stipe. Many of the Tulostomataceae are found in arid sandy regions and *Podaxis* is usually associated with old termite nests. It is probable that wind-driven sand would be a factor in releasing and dispersing the spores.

Some of the genera of the Lycoperdaceae have typical puffball fruitbodies, but in the genera *Geastrum* and *Myriostoma* the exo- and mesoperidium split into stellate segments which become horizontal or recurved and leave the endoperidium (with enclosed gleba) exposed; thus these forms are known as 'earthstars'. In *Geastrum* there is a single apical stoma, but there are several in *Myriostoma*. The typical puffballs include *Lycoperdon*, with an apical stoma; and *Mycenastrum* and *Bovista*, lacking a stoma and distinguished by their capillitium. In *Mycenastrum* the capillitial threads are spiny and free within the peridium; in *Bovista* they branch in a complex dichotomous manner from short central segments.

The peridium of many Lycoperdales dries into a thin flexible membrane which, when hit by splashing raindrops, is depressed momentarily and forces a cloud of spores out through the stoma. One advantage of this liberation mechanism must be that not all the spores are liberated and dispersed simultaneously.

Order Nidulariales

Family Nidulariaceae

The peridium of the Nidulariaceae (Fig. 84) is sessile and globose or funnel-shaped, with one–three layers of wall tissue. At first the fruitbody is entirely closed, but as it matures it opens irregularly or by rupture of a thin apical layer of the peridium termed the **epiphragm.** The gleba rounds off into lenticular segments, the **peridiola,** which contain spores and capillitial threads. In some genera the peridiola are attached to the inner wall of the peridium by means of a complex stalk called a **funiculus.** In the genus *Cyathus* the funiculus consists of a 'purse' containing a very long convoluted

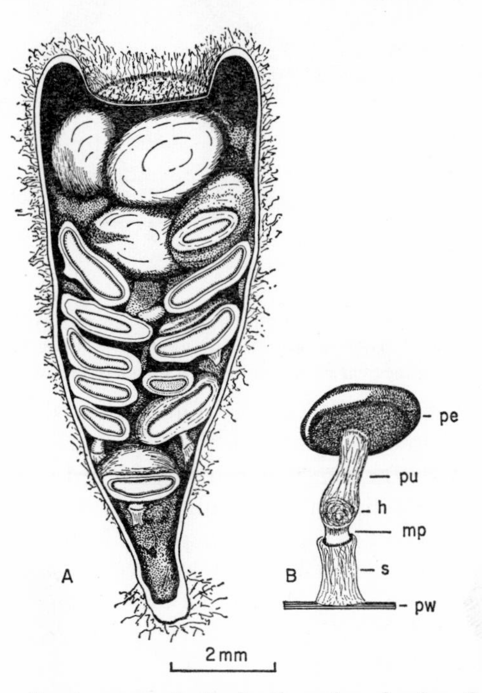

Fig. 84. *Cyathus olla.* A, young fruitbody in section showing the unruptured epiphragm, the peridium and the peridiola; B, detail of the peridiolum (*pe*), purse (*pu*), hapteron (*h*), sheath (*s*), middle piece (*mp*) and wall of the peridium (*pw*). The funiculus comprises *s*, *mp*, and *pu*.

thread ending at the base in a mass of sticky hyphae called the **hapteron,** the whole being inserted in a sheath attached to the peridium. When the fruitbody is open, it forms a cup which can hold water. In wet conditions the sheath surrounding the funiculus becomes gelatinized and the hapteron becomes sticky. Heavy raindrops splash the peridiola out of the peridia,

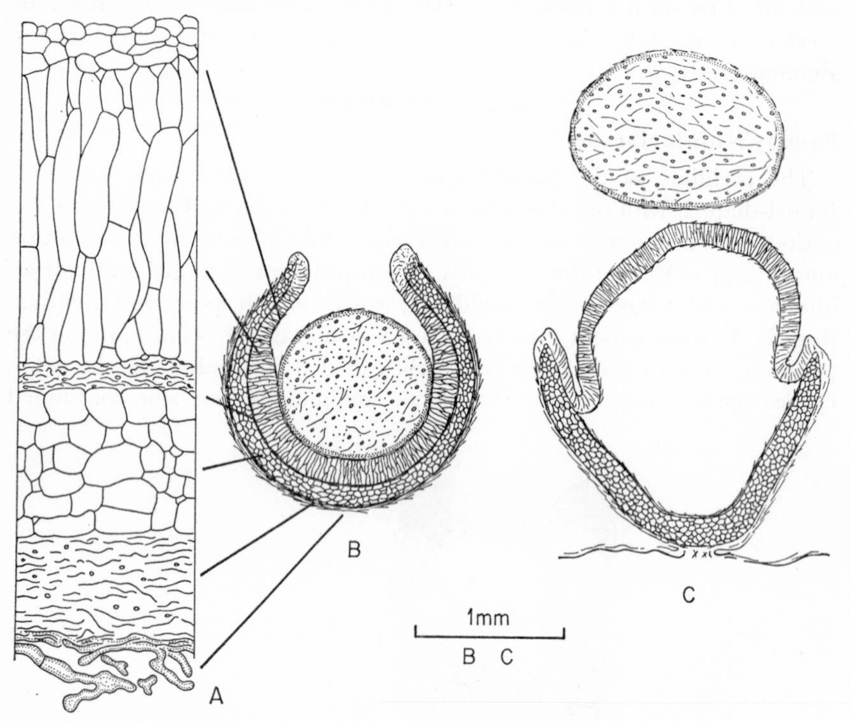

Fig. 85. *Sphaerobolus stellatus* in section. A, detail of the six layers of peridial cells (semidiagrammatic from *camera lucida*); B, peridium containing a single peridiolum or gleba; C, the everted peridium with the gleba thrown out.

rupturing the purse and releasing the funicular cord and hapteron. The sticky disc of the hapteron adheres to vegetation struck by the peridiolum and at the same time the funicular cord is extended to its full length. The momentum of the peridiolum then causes the funicular cord to become twisted around the object to which the hapteron is stuck. In genera where a funiculus is absent, the peridiola themselves are adherent. Some species of these genera are coprophilous, though not exclusively so; thus this method of dispersing the peridiola brings them into a position where they can readily be eaten by browsing herbivores.

Family Sphaerobolaceae

In the family Sphaerobolaceae (Fig. 85) with the genus *Sphaerobolus*, the entire gleba becomes rounded off into a single small peridiolum which lacks a funiculus. The small fruitbodies, 2–3 mm in diameter, are globose and occur partly embedded in wood, dung or soil. The peridium is composed of six well defined layers, of which two are wider and more prominent than the others. The inner of these two is composed of narrow, radially elongated, elastic cells, and the outer of polygonal more rigid cells. The peridium splits from the apex into a number of recurved segments but remains cup-like towards its base. Water collects in the cup surrounding the peridiolum and is absorbed by the layer of elongated, elastic cells. It has also been suggested that osmotic pressure in these cells is increased by transformation of glycogen to sugars. The net effect, in any case, is that the elastic cells stretch while the polygonal cells of the peridium remain unaffected, and the whole elastic layer suddenly everts and throws the peridiolum out with considerable force. Buller (1933) recorded that the peridiolum could be thrown to a maximum vertical distance of 432·5 cm or a horizontal distance of 557·5 cm at an initial velocity of about 33·6 km.p.h.; he calculated that the peridium took 1/1,500 s to evert and that 1/10,000 horse power was required for eversion.

In South Australia, *Sphaerobolus stellatus* occurs not uncommonly on lawns of buffalo grass. The back lawn of most suburban properties is usually the site of a rotary clothesline used for drying laundry. With the aid of modern detergents the whitest and brightest feature of the landscape is undoubtedly the family sheets hanging out to dry. Now unfortunately the peridia of *Sphaerobolus* are positively phototropic. . . . the net result is sheets spotted with dirty black peridiola and worried housewives telephoning research institutes for advice and reassurance.

13

MORPHOLOGICAL SIMILARITIES IN FUNGI

There is so much morphological variety among fungi that it is natural to stress these differences which are to be found, even if only because obvious ones are useful as taxonomic characters. But there are also great similarities among taxonomically unrelated fungi, and it is with these that this chapter is concerned. We shall not be dealing here with strict homology of structures, but only with similarity. Much of what follows is speculative and runs the usual risks of over-generalization; nevertheless, a discussion of this sort may be useful as a reminder that we need a much better knowledge of the properties of hyphae, and their capabilities and limitations in particular situations, if we are to obtain a more unified picture of fungal life. Essentially, what is considered here are some of the potentialities of growth-forms constructed of tubular elements with apical growth, in this case hyphae.

In passing it may be noted that a filamentous system somewhat similar to the hypha is seen in leaf-hairs and root-hairs, but as these are usually extensions of single epidermal cells, and above all are limited in growth, they have little capability for further morphological differentiation. It is of interest, however, to note that root-hairs in dry soil may produce short contorted branches, while those of the same species in wet soil may remain straight, long and unbranched. Well developed filamentous growth systems are found among the algae; the marked similarity in form and structure between certain fungi and algae has already been mentioned briefly.

Reverting now to fungi only, the thesis to be discussed is this: (*a*) hyphae are the basic units of construction of all filamentous fungi, and all other structures (e.g. fruitbodies, sporogenous cells, spores) are derived from them; (*b*) hyphae and their aggregations (e.g. mycelia, stromata, fruitbodies) exist in ecological systems, of which the most prominent are aquatic, epigean, hypogean and parasitic; (*c*) morphological convergence in fungi must be the result of, first, the capabilities and limitations of hyphae (internal factors) and, second, adaptation to function efficiently in similar environments (external factors). Taken together these internal and external factors result in the achievement of similar morphological solutions to problems held in common by different types of fungi. The problems of existence with which fungi are especially concerned are nutrition, growth and survival of the thallus and formation, protection, liberation and dispersal of the spores.

SOME GENERAL ATTRIBUTES OF HYPHAE

There are naturally many exceptions in individual cases but the following general attributes of hyphae assist in explaining some aspects of fungal construction and homoplasy.

Fig. 86, drawn from a culture of *Ascophanus carneus*, illustrates many of these properties. In general hyphae have the ability to: grow apically; lay down septa and thus form cells in a linear row; branch, anastomose, coil, interweave and become short-celled, thus forming simple tissues; become inflated or constricted in parts, especially at the apex of the hypha; become thick-walled and more resistant in parts; cut off specialized short cells which may act as spores themselves, or may give rise to external spores, or to internal spores by protoplasmic cleavage or free-cell formation. These properties are found in various degrees and combinations in most hyphae and will be seen to account for certain structural features common to many different kinds of fungi. The functioning of similar morphological structures, however, depends very largely on nuclear behaviour and may also depend upon other factors about which relatively little is known.

SIMILARITIES IN MICROSCOPIC FEATURES

Apical growth, branching and division of hyphae into a linear succession of cells have the consequence that only rarely do fungal cells divide in three planes, thus true parenchyma is very rare; instead the denser types of tissue are the result of consolidation by branching, interweaving, anastomosis and coiling of hyphae. During this process, septation occurs and the hyphae tend to become short-celled and compressed, resulting in the formation of plectenchyma tissues. Fungal tissues are therefore relatively simple and limited in variety. They differ mainly in the compactness of the hyphae and their degree of division into short cells; also in the pressures tending to make short cells become isodiametric. Sclerotia, rhizomorphs, stromata and fruitbodies may thus be of rather similar final construction although the details of their development may differ.

Septation and anastomosis are undoubtedly connected with the size and strength of thalli. In aquatic lower fungi, water forms a supporting medium,

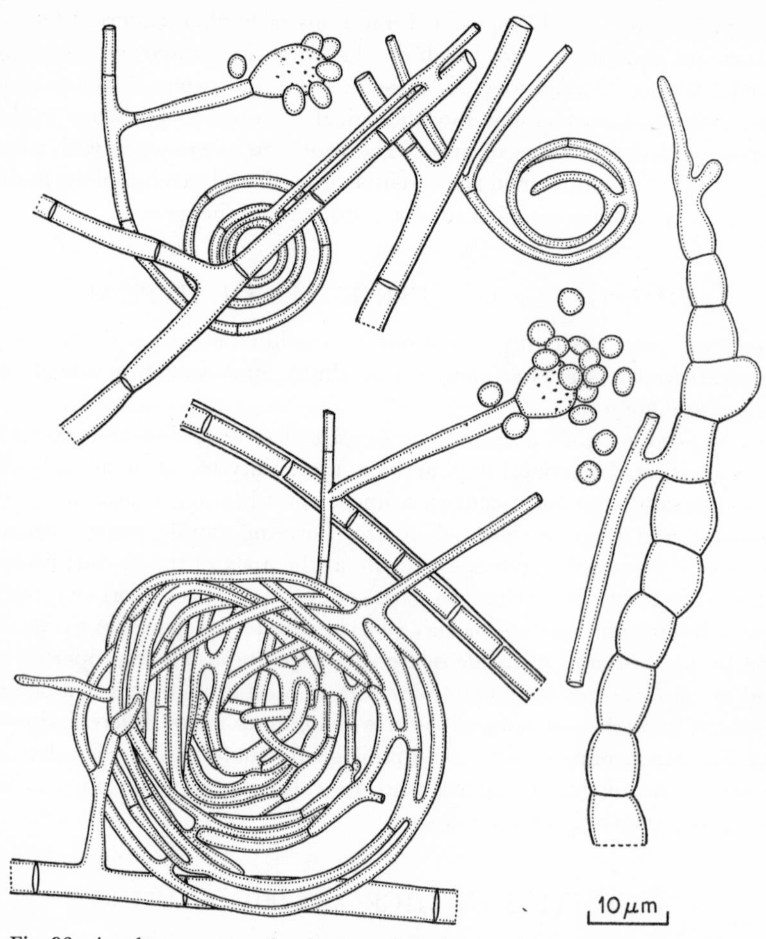

Fig. 86. *Ascophanus carneus*, showing some of the general properties of hyphae. Culture material of the imperfect state, *Rhizoctonia rubiginosa*.

septation of hyphae is irregular and sparse, and somatic anastomosis virtually unknown. The thallus is usually small or composed of a loose mycelium which tends to disrupt easily in the unstable medium. The spores are well adapted to water-dispersal and in general fruitbodies are not formed; simple sporophores are the rule. The majority of Ascomycotina and Basidiomycotina, on the other hand, grow in situations where the spore-bearing structures need to be raised above the substratum for efficient liberation and dispersal of spores and where the developing spores need protection from desiccation. The operation of both these selective pressures has resulted in the formation or relatively large, erect fruitbodies for which some means of support becomes essential. Some of the mechanical support is provided by anastomosis and regular septation of the hyphae; and some, no doubt, is provided by the presence of cementing mucilage which is not washed away in this situation. All hyphae maintain their rigidity partly through the natural strength of their walls, which in many instances do not collapse when empty, and partly by turgor pressure. Some thin-walled hyphae are able to become notably inflated by turgor and play an important part in supporting large fructifications of a soft or fleshy texture which lack skeletal or other strengthening hyphae. Inflated hyphae of this sort are common in the Agaricales, in some clavarioid fungi and in the Phallales; expansion of the developing fruitbody is often dramatically rapid in these. In some of the larger perennial members of the Polyporaceae, the hyphae may be differentiated into skeletal and binding types within the fruitbodies, or may become distinctly thick-walled. All these properties are concerned in the formation of self-supporting fruitbodies with efficient means of spore liberation and dispersal in a non-aquatic environment. Nevertheless, it should not be overlooked that many of the higher fungi may produce conidia on simple sporophores arising from the mycelium, and that the conidia may be quite effective in disseminating the species.

Formation of Spores

Undamaged cells of fragmented mycelia usually have the ability to germinate and grow; this is well known with cultures but it is perhaps not always realized that the same process may occur in nature. Warcup (1959) has picked hyphal fragments from wheat-soil below wilting-point during the hot dry South Australian summer and has found that a large proportion of these are viable though they do not necessarily appear indurated or spore-like. On the other hand, any short cell of a hypha may be a potential thallospore and thus it is to be expected that not only will most thallospores be rather alike in appearance but they will be widely distributed among the groups of fungi.

In the Mastigomycotina and Zygomycotina there is not the great variety of asexual spores that is found in higher fungi, and this may be attributed partly to the relative irregularity of septation in these lower groups: short cells are not only sparse but also tend to be almost devoid of protoplasm, often being formed after its regression; also, most types of spores formed by partition of hyphae are not particularly suited to water-dispersal. Chlamydospores may be formed by members of the lower fungi, especially on media of high osmotic pressure where physiological drought may be assumed, but usually induration and formation of resistant spores is commoner in the higher fungi, as a response to dry conditions. With spores that are formed *de novo* by extrusion of materials from a sporophore apex it is possible that the spore-walls would not be able to harden except under dry conditions. The aquatic Hyphomycetes which immediately contradict this suggestion, are almost certainly derived from terrestrial forms. Resistant sexual spores of the Mastigomycotina are formed often in drying parts of plants or in drying soils.

Where elongation of the hyphal apex is arrested, one type of response is the production of subapical branches. Robertson (1958) has shown that the apices of hyphae of *Fusarium oxysporum* may be made to branch by covering them with water. In this connexion it is interesting to note that the characteristic spores of aquatic Hyphomycetes are most often staurosporous, with long hyphoid arms, and also that the so-called basidiospores of *Digitatispora* (an aquatic member of the Basidiomycotina) are staurosporous. (One may also note that the hairs on the upper surface of the cap of *Acetabularia*, a marine alga, are branched in a similar manner.)

Another response to arrested elongation of the hyphal apex is sometimes the production of spores. As the apex is not only thin-walled and plastic but also the main site of active differentiation and nuclear division, it follows that sporogenous cells and spores are likely to be formed there. Yet the apex can vary in only a limited number of ways: it may inflate, or become narrowed or remain at about its normal width, and within each of these categories the sporogenous cells must all present a rather similar appearance regardless of the type of spore produced. When the apex inflates to form a terminal sporogenous cell, it may give rise to internal spores as in terminal sporangia and asci. On the other hand, the plastic wall of the sporogenous cell may put out extensions which may either round off as asexual spores or act as denticles or sterigmata supporting spores, as with radulaspores (blastospores) and basidiospores. The sporogenous cell may be sac-like and basically similar in all these instances. Their manner of functioning, the type of spore produced and their final appearance after differentiation, all depend upon the pattern of nuclear behaviour in the particular species of fungus.

If the nuclei of the sporogenous cell function asexually, dividing by mitosis, the resulting spores are asexual mitospores. When these are formed simultaneously and usually in large number within a thin-walled inflated sporogenous cell the result is a sporangium and sporangiospores, which may be motile or non-motile. Loss of motility leads from zoospores to aplanospores, and reduction in the number of spores leads to monosporous sporangia which are then conidium-like in position and function. When the contents of the sporogenous cell mature successively, not simultaneously, and the mitospores are successively squeezed out of a narrow apical opening in the sac (corresponding with the exit pore of some sporangia), the result is a phialide with endophialospores. When the apical sporogenous cell is part of the dicaryophase and the dicaryon undergoes karyogamy and meiosis, it results in the production of usually a small number of haploid nuclei which become segregated as meiospores. If these are differentiated within the sac the result is an ascus with ascospores; if they are differentiated externally in extensions of the wall of the sporogenous cell, a basidium and basidiospores result.

When the apical sac forms bud-like extensions of the wall into which nuclei and protoplasm are squeezed, various possibilities are presented. In the asexual cycle the result may be: (*a*) the formation of blastospores — these may be sessile or borne on denticles, and may or may not form a basipetal chain, (*b*) the bud-like extensions on an inflated vesicle may develop into phialides, or metulae and phialides, producing phialospores (e.g. *Aspergillus*). If the process is sexual the result is a basidium, the sterigmata being bud-like extensions of the metabasidium. It is not known in terms of fine structure exactly how far these 'budding' processes correspond, but the analogies are worth further investigation. I suggest that 'budding' of one type or another is a basic property of most hyphae and is not restricted to blastospore formation. Basidiospores are usually regarded as homologous with ascospores, but formed in extensions of the zeugite wall instead of endogenously. Can they not also be regarded as a type of bud-spore formed in limited number after the sexual process in the basidium? Cannot repetitive conidia and basidiospores, or conidia budded from ascospores or basidiospores, each represent a specialized type of budding process? Can a germtube or a branch hypha be regarded as a bud of unlimited growth potential?

In this connexion it is instructive to note that in some of the Lycoperdales and Aphyllophorales (certain species of *Vararia*, *Corticium* and *Peniophora*) and in polypores (*Fomes annosus*), both basidia and oedocephaloid conidiophores may be formed and are sometimes virtually indistinguishable. Oedocephaloid heads of spores are widespread as conidial forms of many Ascomycotina (especially Discomycetes) and fewer Basidiomycotina, and are composed of blastospores arising from denticles which cover a swollen apical vesicle (Fig. 44). In the Zygomycotina (e.g. *Cunninghamella*; Fig. 41, C) they assume the form of reduced sporangia (conidia) borne in a similar manner. In some of the Basidiomycotina mentioned above we can distinguish the spores as basidiospores rather than blastospores only because they are few in number (typically four) and are projected violently from the apices of the short sterigmata. In the Lycoperdales, however, where the fruitbody is closed and the ballistospore function has been lost, the sterigmata are often very much reduced in size and in some instances numerous spores are associated with one basidium and may be formed not only apically but also laterally on the basidium (*see*, e.g. Bessey, 1950, Fig. 187 B, for basidia of *Calostoma*). Are these, then, basidiospores or conidia? Only cytology can give the answer, but this does seem to suggest that the basidium can be regarded as a specialized type of conidiophore — specialized because it has a sexual pattern of nuclear behaviour; morphologically the two may be virtually identical. I suggest that phialides, capitate vesicular conidiophores, asci and basidia are all evolutionary developments from terminal sporangia; it is their cytological patterns that make them analogous and not necessarily homologous organs. In this view, the ascus can be regarded as a sporangium with a sexual function (or more precisely a zygosporangium); the basidium as a conidiophore with a sexual function; and the conidiophore or phialide as a development from a sporangium. But all these are derived ultimately from the plastic apex of a hypha with all its capabilities for differentiation. This may prove to be an interesting line of enquiry, as it is of speculation. We tend to regard unconnected asexual states of fungi as possibly representing species which have lost their sexual capacity, but is it not possible that some may never have been sexual organisms? Against this may be argued that sex is so well established and so successful even in the simplest types of fungi that its complete absence in some higher fungi cannot reasonably be expected except as a reduction process.

Proliferation, Croziers and Clamps

Renewed growth and development as an offshoot from a part of the thallus whose growth has been arrested, or from a cell adjoining an exhausted sporogenous cell, is known as **proliferation.** Two main types of proliferation occur, with examples to be found in all Subdivisions of the Eumycota. Distal or terminal proliferation takes place by renewed growth from a terminal cell, or from a penultimate cell, through the empty walls of the exhausted cell. The exhausted cell may be a somatic hyphal cell (e.g. in *Rhizoctonia solani*; *Endogone*; fragmented hyphae), or a sporangium (*Saprolegnia*), an ascus (*Ascoidea*), a basidium (*Repetobasidium*; *Galzinia*), or an annellophore (many Deuteromycotina). Lateral or sympodial proliferation, on the other hand, takes place by a new branch or sporogenous cell being formed from the penultimate cell from beneath the septum at the base of the exhausted apical cell; again there are examples concerning somatic hyphae, sporangia (*Achlya*), asci, basidia (many Aphyllophorales) and sporophores of Deuteromycotina.

Both croziers and clamp connexions are able to act as proliferative structures. With croziers, proliferation occurs in two ways, sometimes both in the same species: (*a*) The penultimate dicaryotic cell of the crozier may, instead of becoming an ascus mother cell, elongate and divide into a chain of cells with the formation of another crozier at the apex, and so on in a series. Usually an ascus terminates the series but this is not necessarily so; some chains remain sterile. (*b*) The basal cell of the crozier unites with its terminal cell to form a common dicaryotic cell which elongates laterally and forms another crozier. In most instances, clamped hyphae in the Basidiomycotina correspond with the first of these types of proliferation and the basidium terminates a long series of clamped, sterile segments of hypha. In a few instances proliferation of the second type is known, where this occurs laterally from the clamp at the base of a basidium.

In drawing attention to the essentially similar methods of proliferation of clamps and croziers, Rogers (1936) has also reviewed hypotheses as to the ways in which these structures may have originated. The prototype crozier he postulates to have arisen as follows. In the narrowed apex of an ascogenous hypha the dicaryotic nuclei, A and B, might well have to lie one above the other instead of side by side, because of limitations of space. Thus on conjugate division the nuclear spindles would lie longitudinally one above the other, and the nuclei A and A would lie above the other pair, B and B. With septum-formation across the spindles, one nucleus A would occupy the apical cell, a pair of nuclei A and B the penultimate cell, and one nucleus B the third cell. The penultimate cell with its pair of complementary nuclei would enlarge into an ascus, pushing the apical cell sideways and bending it over. It would then be likely to anastomose with the third cell, whose nucleus complements its own. Essentially, a crozier would have been formed. Once the pattern had become established it would be possible in more advanced Ascomycotina for curving of the apex to take place before, rather than after, conjugate division.

In Basidiomycotina, which are held to be derived from Ascomycotina, it is suggested that the dicaryotic hyphal apex could have elongated beyond the region of recurvature before conjugate division, thus bringing about the lateral rather than the terminal position of the clamp. Even though clamps occur on what appear to be purely somatic hyphae of Basidiomycotina, these do in fact ultimately bear the zeugites just as the ascogenous hyphae do. Rogers (1936) presents extensive arguments in favour of the homology of the crozier and clamp connexion, and hence in favour of the hypothesis that prototype Basidiomycotina would have been clamp-bearing and derived from crozier-bearing Ascomycotina; homologous structures of this sort, with their complicated cytology, could hardly have arisen repeatedly during the course of evolution. Two points are worth noting in this regard. First, whatever may have been the position in prototypes, both croziers and clamps in modern fungi may be associated with wide hyphae; indeed the ascogenous hyphae are often very wide at their apices. Second, hyphae tend to have the same potentialities: given a narrow dicaryotic hypha, whether of Ascomycotina or Basidiomycotina, the same hyphal properties, nuclear potentialities and spatial relationships would operate, and thus the crozier or clamp could very well have evolved repeatedly in different groups of primitive dicaryophytes. The crux of the matter is not so much the origin of croziers and clamps, but rather the origin of dicaryotism and conjugate division. To my mind a weightier argument for deriving Basidiomycotina from Ascomycotina is to be found in the fact that in Ascomycotina the dicaryophase usually occupies a smaller and more dependent phase of the life-cycle than in Basidiomycotina, and the rise of the dicaryophase in the latter has been accompanied by marked decline in overt sexuality — features which are seen in transition in the Ascomycotina.

Coiling of Hyphae

The ability of hyphae to coil is perhaps not worthy of much emphasis, but it is concerned in the formation of helicospores (which are distinctly hyphoid in appearance), in some types of snares of predacious Hyphomycetes, and sometimes in the formation of sclerotia, stromata and spore-balls. The antibiotic griseofulvin, originally isolated from *Penicillium griseofulvum*, produces marked curling of hyphae.

Anastomosis and Plasmogamy

Some of the effects of anastomosis in somatic hyphae have already been noted. It is profitable, however, to consider all forms of plasmogamy as types of anastomosis — sexual anastomosis. Only five general methods of plasmogamy are known in fungi and it is indeed difficult to visualize other possible ways of bringing protoplasts, gametes or nuclei together in one cell. Many striking similarities can be seen in the conjugating cells of fungi in different major taxa, and also in filamentous algae, no doubt because of the limited means of anastomosis. When two cells meet and fuse it is usual for the common cell so formed to become arched or looped above the general level of the hyphae bearing the conjugating cells. The arched cell then continues to elongate and differentiate; its base is often bifurcate and reminiscent of the remains of a crozier at the base of an ascus. The suggestion has been made, particularly with reference to *Dipodascus*, that an ascus is the homologue of a zygosporangium.

Anastomosis frequently takes place between cells of hyphae of the same species, but may also occur between branches or adjacent cells of a single hypha. In *Ancylistes* self-anastomosis of adjacent cells is effected by a looped bypass superficially very like a clamp connexion but without similarity in cytology. However, both croziers and clamp connexions are formed by self-anastomosis and with strikingly similar cytological processes.

The type of sexual anastomosis becomes simpler and less varied, eventually becoming essentially somatic, as one passes in review from the lower fungi to the Ascomycotina and finally to the Basidiomycotina. The variety of sexual morphological structures and the variety of methods of conjugation become much reduced. Conjugation of motile gametes gives place to conjugation of gametangia, and finally to anastomosis of somatic hyphae. Nuclear fusion takes place in the gametangia or in cells derived from the dicaryotic hyphae: thus the function of the gamete tends to be assumed by the gamete mother cell. This type of reduction, which finds a parallel in reduction of the asexual processes in fungi, is often regarded as degenerate, but has nevertheless been highly successful in fungi. With a decline in sexuality there has been a trend towards the production of dicaryophytes, fruitbodies of higher fungi, which have been a major factor in adaptation to existence on land.

SIMILARITIES IN MACROSCOPIC STRUCTURES

The effect of apical centrifugal growth of hyphae in producing a radial growth pattern in a filamentous thallus has been noted previously.

In gross morphology the main types of fructifications may take the form of: resupinate stromata, cushions, cups, brackets, clubs, stipitate pilei, flask-like conceptacles (pycnidia, perithecia, spermogonia), layer-like stromata (acervuli, sori) and closed structures (cleistocarps). Most of these forms are widely encountered among the higher fungi and their conidial states, but it should be noted that the lower fungi are not well represented. The few lower fungi that form a very limited range of fructifications are either hypogean (rudimentary cleistocarps in *Endogone*) or plant parasites (sori in *Physoderma*; *Albugo*). Simple sporophores are the rule in aquatic lower fungi.

In lower fungi whose thalli are supported by water, various factors operate to keep the thallus loosely constructed and small: the hyphae tend to be kept apart by water and are generally considered to be incapable of somatic anastomosis; there are relatively few septa to strengthen the hyphae; water is not static and its movements would tend to fragment thalli; nutrients are brought to the hyphae by water, thus there is less pressure on them to send out extensive systems of foraging branches. All these factors tend to prohibit the formation of stromata and the sporophores remain simple and relatively unspecialized yet well suited to the types of spores produced. Aquatic lower fungi have motile spores or thallospores or resting spores with uncomplicated mechanisms of liberation and dispersal; virtually any type of cell would be capable of releasing gametes or zoospores into water and they would be effectively dispersed. The danger of seasonal desiccation would not appear to be a great problem since the sexual spores and thallospores are resistant, while in diplanetic or multiplanetic species the zoospores are able to encyst between their periods of motility.

The mycelium of higher fungi in an epigean habitat is usually very extensive, closely interwoven, regularly septate, able to anastomose and coalesce into hyphal strands, stromata, sclerotia and fruitbodies with co-ordinated growth. The thallus is continually added to, not only because it has to forage for nutrient in wood or soil, but also because these are relatively static media without movement which would disrupt the thallus. Fungi in such a habitat require larger, compound sporophores to support the spore-bearing cells above the substratum in such a position that anemochory and entomochory can operate. This support is achieved by interweaving and anastomosis of hyphae to form tissues, or by differentiation of inflatable, skeletal and binding types of hyphae, and also by the presence of regular septa. Though little is known about translocation in fungi, it must be assumed that such stromata and fruitbodies translocate efficiently. In higher fungi the danger of desiccation is reduced by the bulk of the fruitbody protecting the spores; also resistant thallospores are not uncommon and pigmentation and thickening of spores and hyphae may also affect survival. Since, broadly speaking, the hyphae of higher fungi have many of the same capabilities and limitations, and the mechanical principles in building fructifications are the same, the fructifications tend to assume a limited range of homoplastic forms which may differ in internal construction and development but most notably in the refinements of construction associated with particular types of spore-release and dispersal. Highly convergent forms of hymenial configuration also occur in unrelated groups of Basidiomycotina and Ascomycotina; they are the expression of economical ways of increasing the spore-bearing area without unduly increasing the bulk and hence the food requirements. In both these subdivisions lines of phylogenetic development from gymnocarpy, through hemiangiocarpy, to angiocarpy, are evident.

With hypogean fungi there is a tendency for the fruitbodies to become rounded and closed, often with solid internal tissues, and often with rounded sporogenous cells or spores. These features are probably connected with the approximately equal pressures produced as the developing fructifications expand in the surrounding soil. Rounded hypogean fructifications are seen in at least one member of the lower fungi (*Endogone*), in various Ascomycotina (particularly Tuberales) and in some Basidiomycotina (Hymenogastrales; 'eggs' of Phallales); some of these may be remarkably similar in external morphology.

Those fungi that are internal parasites of plants and animals are usually reduced in size, sometimes occupying single host cells. Reduction in size may be accompanied by simplification of the thallus into yeast-like cells or into hyphae segmented into 'hyphal bodies'. As with the hypogean habitat, equal pressures around developing cells tend to produce a rounded thallus; the thallus has a granular aspect in some of the fungi causing internal disease of humans, e.g. *Histoplasma* and *Cryptococcus*.

In xerophytic situations the mycelium is often pigmented, radiating and appressed, and anchoring structures such as appressoria or short lateral branches such as hyphopodia may be common. Such fungi grow best in humid regions, since the substratum is physiologically dry. Many of the epiphyllous asterinoid Ascomycotina, the Laboulbeniales on insects and *Piedraia* on hair, come into this category.

There are obviously other ecological sites that have not been considered and in which particular morphological potentialities of fungi become expressed.

I am conscious of the liberties I have taken in this chapter: repetition of material from other chapters, speculation, ignoring of strict homology, and above all over-simplification of profound mysteries. But my aim has been achieved if a glimpse has been given of fungi as plastic and highly successful organisms, occupying various environments and responding to these in a number of ways, yet on the whole presenting a picture of unity in diversity.

REFERENCES

AINSWORTH, G. C. (1966). A general purpose classification of fungi. *Bibliography of Systematic Mycology*, 1966 1–4 (Commonwealth Mycological Institute, Kew).

AINSWORTH, G. C. and SAMPSON, KATHLEEN (1950). *The British Smut Fungi* (*Ustilaginales*) (Commonwealth Mycological Institute, Kew).

ALEXOPOULOS, C. J. (1962). *Introductory Mycology*, 2nd ed. (Wiley, New York).

ARBER, AGNES (1953). *Herbals*, 2nd ed. (Cambridge University Press).

ARONSON, J. M. (1966). The cell wall, in *The Fungi* edit. by G. C. Ainsworth and A. S. Sussman, vol. 2 (Academic Press, New York).

ARX, J. A. VON and MÜLLER, E. (1954). *Die Gattungen der amerosporen Pyrenomyceten* (Büchler, Bern).

BARNES, B. (1928). Variations in *Eurotium herbariorum* (Wigg.) Link, induced by the action of high temperatures. *Ann. Bot.*, **42**, 783.

BARNES, B. (1930). Variations in *Botrytis cinerea* Pers., induced by the action of high temperatures. *Ann. Bot.*, **44**, 825.

BATRA, L. R. (1963). Ecology of ambrosia fungi and their dissemination by beetles. *Trans. Kansas Acad. Sci.*, **66**, 213.

BATRA, L. R. (1966). Ambrosia fungi: extent of specificity to ambrosia beetles. *Science, N.Y.*, **153**, 193.

BATRA, L. R. (1967). Ambrosia fungi: A taxonomic revision, and nutritional studies of some species. *Mycologia*, **59**, 976.

BENEKE, E. S. (1963). Calvatia, calvacin and cancer. *Mycologia*, **55**, 257.

BENJAMIN, C. R. (1955). Ascocarps of *Aspergillus* and *Penicillium*. *Mycologia*, **47**, 669.

BENJAMIN, R. K. (1959). The merosporangiferous Mucorales. *El Aliso*, **4**, 321.

BENJAMIN, R. K. (1966). The merosporangium. *Mycologia*, **58**, 1.

BESSEY, E. A. (1950). *Morphology and Taxonomy of Fungi* (Constable, London).

BISBY, G. R. (1953). *Introduction to Taxonomy and Nomenclature of Fungi*, 2nd ed. (Commonwealth Mycological Institute, Kew).

BLAKESLEE, A. F. (1904). Sexual reproduction in the Mucorineae. *Proc. Am. Acad. Arts Sci.*, **40**, 205.

BLUMER, S. (1933). *Die Erysiphaceen Mitteleuropas.* (Fretz, Zurich).

BOIDIN, J. (1958). *Essai biotaxonomique sur les Hydnés résupinés et les Corticiés.* Thèse, Univ. de Lyon 1954, No. 202.

BONNER, J. T. (1959). *The Cellular Slime Molds* (Princeton University Press, Princeton).

BOOTH, C. (1959). Studies of Pyrenomycetes IV. *Nectria* (Part 1). *Mycol. Pap.*, No. 73 (Commonwealth Mycological Institute, Kew).

BOURCHIER, R. J. (1957). Variation in cultural conditions and its effect on hyphal fusion in *Corticium vellereum. Mycologia*, **49**, 20.

BRACKER, C. E. (1967). Ultrastructure of fungi. *Ann. Rev. Phytopath.*, **5**, 343.

BROWN, W. (1915). Studies in the physiology of parasitism I. The action of *Botrytis cinerea. Ann. Bot.*, **29**, 313.

BROWN, W. (1916). Studies in the physiology of parasitism III. On the relation between the 'infection drop' and the underlying host tissue. *Ann. Bot.*, **30**, 399.

BROWN, W. (1917). Studies in the physiology of parasitism IV. On the distribution of cytase in cultures of *Botrytis cinerea. Ann. Bot.*, **31**, 489.

BULLER, A. H. R. (1909, 22, 33, 34,). *Researches on Fungi*, **1, 2, 5, 6** (Longmans, Green, London).

BULMER, G. S. (1964). Spore germination of forty-two species of puffballs. *Mycologia*, **56**, 630.

BURGES, A. (1955). Problems associated with the species concept in mycology. In *Species studies in The British Flora* edit. by J. E. Lousley, p. 65.

BUTLER, GILLIAN M. (1957). The development and behaviour of mycelial strands in *Merulius lacrymans* (Wulf.) Fr. I. Strand formation during growth from a food-base through a non-nutrient medium. *Ann. Bot., N.S.*, **21**, 523.

BUTLER, GILLIAN M. (1958). The development and behaviour of mycelial strands in *Merulius lacrymans* (Wulf.) Fr. II. Hyphal behaviour during strand formation. *Ann. Bot., N.S.*, **22**, 219.

CARLISLE, M. J., LEWIS, B. G. MORDUE, ELIZABETH M. and NORTHOVER, J. (1961). The development of coremia. I. *Penicillium claviforme. Trans. Br. mycol. Soc.*, **44**, 129.

CARTER, M. V. and MOLLER, W. J. (1961). Factors affecting the survival and dissemination of *Mycosphaerella pinodes* (Berk. & Blox.) Vesterg. in South Australian irrigated pea fields. *Aust. J. Agric. Res.*, **12**, 878.

CHILVERS, G. A. and PRIOR, L. D. (1965). The structure of Eucalypt mycorrhizas. *Aust. J. Bot.*, **13**, 245.

CHRISTENSEN, C. M. (1961). *The Molds and Man* (University of Minneapolis Press, Minneapolis).

CLARE, B. G. (1964). Erysiphaceae of South-Eastern Queensland. *Univ. Qld. Papers*, **4** (10), 111 (University of Queensland Press, St. Lucia).

CLELAND, J. B. (1934). *Toadstools and Mushrooms and other larger Fungi of South Australia.* Part 1 (Government Printer, Adelaide).

CLERK, G. C. and AYESU-OFFEI, E. N. (1967). Conidia and conidial germination in *Leveillula taurica* (Lév.) Arn. *Ann. Bot., N.S.*, **31**, 749.

COOK, R. J. and SNYDER, W. C. (1965). Influence of host exudates on growth and survival of germlings of *Fusarium solani f. phaseoli* in soil. *Phytopathology*, **55**, 1021.

CORNER, E. J. H. (1929). Studies in the morphology of Discomycetes I. The marginal growth of apothecia. II. The structure and development of the ascocarp. *Trans. Br. mycol. Soc.*, **14**, 263.

CORNER, E. J. H. (1932*a*). A *Fomes* with two systems of hyphae. *Trans. Br. mycol. Soc.*, **17**, 51.

CORNER, E. J. H. (1932*b*). The fruiting body of *Polystictus xanthopus* Fr. *Ann. Bot.*, **46**, 71.

CORNER, E. J. H. (1950). A Monograph of *Clavaria* and Allied Genera. (Oxford University Press).

CORNER, E. J. H. (1954). The classification of the higher fungi. *Proc. Linn. Soc. Lond.*, Session 165, Part 1, pp. 4–6.

CORNER, E. J. H. (1959). The importance of tropical taxonomy to modern botany. *Gardens Bull. Singapore*, **17**, 209.

CORNER, E. J. H. (1968). *The Life of Plants* (Mentor Books, New American Library, New York and Toronto).

COUCH, J. N. (1938). *The Genus Septobasidium* (University of North Carolina Press, Chapel Hill).

CUNNINGHAM, G. H. (1944). *The Gasteromycetes of Australia and New Zealand* (Dunedin).

DENNIS, R. W. G. (1949). A revision of the British Hyaloscyphaceae with notes on related European species. *C.M.I. Mycological Papers* No. 32 (Commonwealth Mycological Institute, Kew).

DENNIS, R. W. G. (1956). A revision of the British Helotiaceae in the herbarium of the Royal Botanic Gardens, Kew, with notes on related European species. *C.M.I. Mycological Papers* No. 62 (Commonwealth Mycological institute, Kew).

DENNIS R. W. G. (1960). *British Cup Fungi and their Allies* (Ray Society, London).

DOIDGE, ETHEL M. (1942). A revision of the South African Microthyriaceae *Bothalia*, **4**, 273–420.

DOMINIK, T. (1959). Synopsis of a new classification of the ectotrophic mycorrhizae established on morphological and anatomical characteristics. *Mycopath. Mycol. appl.*, **11**, 359.

DONK, M. A. (1931). Revisie van de Nederlandse Heterobasidiomycetae en Homobasidiomycetae-Aphyllophoraceae I. *Nederl. Mycol. Vereen. Med.*, **18, 19, 20,** 67–200.

DONK, M. A. (1954). A note on sterigmata in general. *Bothalia*, **6**, 301.

DONK, M. A. (1956). Notes on resupinate Hymenomycetes II. The tulasnelloid fungi. *Reinwardtia*, **3**, 363.

DONK, M. A. (1964). A conspectus of the families of Aphyllophorales. *Persoonia*, **3**, 199.

FENNELL, DOROTHY I. and WARCUP, J. H. (1959). The ascocarps of *Aspergillus alliaceus*. *Mycologia*, **51**, 409.

FENNER, E. ALINE (1932). *Mycotypha microspora*, a new genus of the Mucoraceae. *Mycologia*, **24**, 187.

FINDLAY, W. P. K. (1951). The development of *Armillaria mellea* rhizomorphs in a water tunnel. *Trans. Br. mycol. Soc.*, **34**, 146.

FLENTJE, N. T. and STRETTON, HELENA M. (1964). Mechanisms of variation in *Thanatephorus cucumeris* and *T. praticolus*. *Aust. J. biol. Sci.*, **17**, 686.

FLENTJE, N. T., STRETTON, HELENA M. and HAWN, E. J. (1963). Nuclear distribution and behaviour throughout the life cycles of *Thanatephorus*, *Waitea* and *Ceratobasidium* species. *Aust. J. biol. Sci.*, **16**, 450.

FRANCKE-GROSMAN, H. (1967). Ectosymbiosis in wood-inhabiting insects. In *Symbiosis* II, edit. by S. M. Henry (Academic Press, New York and London).

FRANK, B. (1883). Über einige neue u. weniger bekannte Pflanzenkrankheiten. *Ber. Dt. bot. Ges.*, **1**, 62.

FRAYMOUTH, JOAN (1956). Haustoria of the Peronosporales. *Trans. Br. mycol. Soc.*, **39**, 79.

FRIES, E. M. (1874). *Hymenomycetes Europaei sive epicriseos systematis mycologici* (Uppsala).

GARRETT, S. D. (1963). *Soil fungi and soil fertility* (Pergamon Press, Oxford; Macmillan, New York).

GILBERT, H. C. (1935). Critical events in the life history of *Ceratiomyxa*. *Am. J. Bot.*, **22**, 52.

GILKEY, HELEN M. (1939). *Tuberales of North America*. Oregon State Monographic Studies in Botany, No. 1 (Oregon State College, Corvallis).

GILMOUR, J. S. L. (1961). Taxonomy. In *Contemporary Botanical Thought*, edit. by A. M. MacLeod and L. S. Cobley (Oliver and Boyd, Edinburgh).

GOLOVIN, P. N. (1958). The idea of species in mycology. In *The Species Problem in Botany*. Acad. Sci. U.S.S.R., Moscow, 1958 (Translation 65989, Canada Dept. Agriculture).

GREGORY, P. H. (1966). The fungus spore: what it is and what it does. In *The Fungus Spore*, edit. by M. F. Madelin (Butterworths, London).

GRIFFIN, D. M. (1963). Soil moisture and the ecology of soil fungi. *Biol. Rev.*, **38**, 141.

HANSFORD, C. G. (1946). The foliicolous Ascomycetes, their parasites and associated fungi. *C.M.I. Mycological Paper*, No. 15 (Commonwealth Mycological Institute, Kew).

Harley, J. L. (1959). *The Biology of Mycorrhiza* (Leonard Hill, London).

Hawker, Lilian E. (1957). The Physiology of Reproduction in Fungi (Cambridge University Press).

Hawker, Lilian E. (1965). Fine structure of fungi as revealed by electron microscopy. *Biol. Rev.*, **40**, 52.

Heim, R. (1948). Phylogeny and natural classification of macro-fungi. *Trans. Br. mycol. Soc.*, **30**, 161.

Heslop-Harrison, J. (1953). *New concepts in flowering-plant taxonomy* (Heinemann, London).

Hesseltine, C. W. (1965). A millennium of fungi, food, and fermentation. *Mycologia*, **57**, 149.

Hochreutiner, B. P. G. (1929). Sur la systématique en général et celle des columniferes en particulier. *Verhand. schweiz. nat. Gesell.*, (2), 151.

Hughes, S. J. (1953). Conidiophores, conidia and classification. *Can. J. Bot.*, **31**, 577.

Ingold, C. T. (1942). Aquatic Hyphomycetes of decaying alder leaves. *Trans. Br. mycol. Soc.*, **25**, 339.

Ingold, C. T. (1961). *The Biology of Fungi* (Hutchinson, London).

Ingold, C. T. (1965). *Spore Liberation* (Oxford University Press).

Ingold, C. T. and Zoberi, M. H. (1963). The asexual apparatus of Mucorales in relation to spore liberation. *Trans. Br. mycol. Soc.*, **46**, 115.

Johnson, T. W. and Sparrow, F. K. (1961). *Fungi in Oceans and Estuaries* (J. Cramer, Weinheim).

Karling, J. S. (1932). Studies in the Chytridiales. VII. The organization of the chytrid thallus. *Am. J. Bot.*, **19**, 41.

Kendrick, W. B. and Weresub, Luella K. (1966). Attempting neo-Adansonian computer taxonomy at the ordinary level in Basidiomycetes. *Syst. Zool.*, **15**, 307.

Kühn, J. C. (1858). *Die Krankheiten der Kulturgewächse, ihre ursachen und ihre Verhütung* (Besselmann, Berlin).

Langeron, M. (1945). *Précis de Mycologie* (Masson et Cie, Paris).

Large, E. C. (1940). *The Advance of the Fungi* (Jonathan Cape, London).

Lawrence, G. H. M. (1951). Taxonomy of vascular plants (Macmillan, New York).

Lemke, P. A. (1964). The genus *Aleurodiscus* (sensu stricto) in North America. *Can. J. Bot.*, **42**, 213.

Linder, D. H. (1943). The genera *Kickxella*, *Martensella*, and *Coemansia*. *Farlowia*, **1**, 49.

Lloyd, C. G. (1902). The genera of Gastromycetes. In *Mycological Writings*, **1**.

Lloyd, C. G. (1913). *Mycological Writings*, **4**, Letter 48, 12.

Lentz, P. L. (1954). Modified hyphae of Hymenomycetes. *Bot. Rev.*, **20**, 135.

Lowy, B. (1968). Taxonomic problems in the Heterobasidiomycetes. *Taxon*, **17**, 118.

LUTTRELL, E. S. (1951). Taxonomy of the Pyrenomycetes. *University of Missouri Studies*, **24** (3), 1.

LUTTRELL, E. S. (1955). The ascostromatic Ascomycetes. *Mycologia*, **47**, 511.

LUTTRELL, E. S. (1958). The function of taxonomy in mycology. *Mycologia*, **50**, 942.

LUTTRELL, E. S. (1965). Classification of the Loculoascomycetes. *Phytopathology*, **55**, 828.

MADELIN, M. F. (1966). The genesis of spores of higher fungi. In *The Fungus Spore*, edit. by M. F. Madelin (Butterworths, London).

MANIER, J. F. and LICHTWARDT, R. W. (1968). Révision de la Systématique des Trichomycètes. *Ann. Sci. Nat. Bot., Paris* ser. 12, **9**, 519.

MARTIN, G. W. (1932). Systematic position of the slime-moulds and its bearing on the classification of fungi. *Bot. Gaz.*, **93**, 421.

MARTIN, G. W. (1938). The morphology of the basidium. *Am. J. Bot.*, **25**, 68.

MARTIN, G. W. (1955). Are fungi plants? *Mycologia*, **47**, 779.

MARTIN, G. W. (1957). The tulasnelloid fungi and their bearing on basidial terminology. *Brittonia*, **9**, 25.

MARTIN, G. W. (1960). The systematic position of the Myxomycetes. *Mycologia*, **52**, 119.

MASON, E. W. (1933). Annotated account of fungi received at the Imperial Mycological Institute. List 2, fasc. 2, 1–67 (I.M.I., Kew).

MASON, E. W. (1937). Annotated account of fungi received at the Imperial Mycological Institute. List 2, fasc. 3 (general part), 68–99 (I.M.I., Kew).

MASON, E. W. (1940). On specimens, species and names. *Trans. Br. mycol. Soc.*, **24**, 115.

MASON, E. W. (1941). Annotated account of fungi received at the Imperial Mycological Institute. List 2, fasc. 3 (special part), 100–144 (I.M.I., Kew).

MASON, E. W. (1948). Some common mould. *The Naturalist* (Yorkshire), Jan.–March, 5.

MEEUSE, A. D. J. (1964). A critique of numerical taxonomy. Systematics Association, publ. No. 6: 115 (The Systematics Association, London).

MOUNCE, IRENE and MACRAE, RUTH (1938). Interfertility phenomena in *Fomes pinicola*. *Can. J. Res.* (C), **16**, 354.

MÜLLER, E. and ARX, J. A. VON (1962). *Die Gattungen der didymosporen Pyrenomyceten* (Büchler, Wabern-Bern).

MUNK, A. (1962). An approach to an analysis of taxonomic method with main reference to higher fungi. *Taxon*, **11**, 185.

NANNFELDT, J. A. (1932). *Studien über die morphologie und systematic der nicht-lichenisierten Inoperculaten Discomyceten* (Almqvist and Wiksells, Uppsala).

NEUHOFF, W. (1924). Zytologie und systematische Stellung der Auriculariaceen und Tremellaceen. *Bot. Arkin.*, **8**, 250.

NOBLES, MILDRED K. (1948). Studies in forest pathology VI. Identification of cultures of wood-rotting fungi. *Can. J. Res.* (C), **21**, 281.

NOBLES, MILDRED K. (1958). Cultural characters as a guide to the taxonomy and phylogeny of the Polyporaceae. *Can. J. Bot.*, **36**, 883.

OLIVE, L. S. (1964). Spore discharge mechanism in Basidiomycetes. *Science, N.Y.*, **146**, 542.

PANTIDOU, MARIA E. (1961). Cultural studies of Boletaceae. *Can. J. Bot.*, **39**, 1149.

PARBERY, D. G. (1969). *Amorphotheca resinae* gen. nov., sp. nov.: the perfect state of *Cladosporium resinae*. *Austr. J. Bot.*, **17**, 331.

PATOUILLARD, N. (1900). Essai taxonomique sur les familles et les genres des Hyménomycetès (Lucien, Lons-le-Saunier).

PONTECORVO, G. and ROPER, J. A. (1952). Genetic analysis without sexual reproduction by means of polyploidy in *Aspergillus nidulans*. *J. gen. Microbiol.*, **6**, vii abstract.

PRINGLE, R. B. and SCHEFFER, R. P. (1964). Host-specific plant toxins. *Ann. Rev. Phytopath.*, **2**, 133.

RAMSBOTTOM, J. (1937). Dry rot in ships. *Essex Naturalist*, **25**, 231.

REICHLE, R. E. and ALEXANDER, J. V. (1965). Multiperforate septations, Woronin bodies, and septal plugs in *Fusarium*. *J. Cell. Biol.*, **24**, 489.

REMSBERG, RUTH E. (1940). Studies in the genus *Typhula*. *Mycologia*, **32**, 52.

RISHBETH, J. (1963). Stump protection against *Fomes annosus* III. Inoculation with *Peniophora gigantea*. *Ann. Appl. Biol.*, **52**, 63.

ROBERTSON, N. F. (1958). Observations on the effects of water on the hyphal apices of *Fusarium oxysporum*. *Ann. Bot., N.S.*, **22**, 159.

ROBERTSON, N. F. (1965). The fungal hypha. *Trans. Br. mycol. Soc.*, **48**, 1.

ROBINOW, C. F. (1957*a*). The structure and behaviour of the nuclei in spores and growing hyphae of Mucorales I. *Mucor hiemalis* and *Mucor fragilis*. *Can. J. Microbiol.*, **3**, 771.

ROBINOW, C. F. (1957*b*). The structure and behaviour of the nuclei in spores and growing hyphae of Mucorales II. *Phycomyces blakesleeanus*. *Can. J. Microbiol.*, **3**, 791.

ROBINOW, C. F. (1963). Observations on cell growth, mitosis, and division in the fungus *Basidiobolus ranarum*. *J. Cell Biol.*, **17**, 123.

ROGERS, D. P. (1934). The basidium. *University of Iowa Studies in Natural History*, **16**, 160.

ROGERS, D. P. (1936). Basidial proliferation through clamp-formation in a new *Sebacina*. *Mycologia*, **28**, 347.

ROGERS, D. P. (1958). The philosophy of taxonomy. *Mycologia*, **50**, 326.

SACCARDO, P. A. (1886, 1887, 1888). *Sylloge fungorum* **4**, **5**, **6** (Patavia).

SALMON, E. S. (1900). A monograph of the Erysiphaceae. *Mem. Torrey bot. Cl.*, **9**, 1.

SAVILE, D. B. O. (1965). Spore discharge in Basidiomycetes: A unified theory. *Science*, **147**, 165.

SINGER, R. (1962a). *The Agaricales in Modern Taxonomy*. 2nd ed. (J. Cramer, Weinheim).

SINGER, R. (1962b). *Keys for the Determination of the Agaricales* (J. Cramer, Weinheim).

SMITH, A. H. (1966). The hyphal structure of the basidiocarp. In *The Fungi II*, edit. by G. C. Ainsworth and A. S. Sussman (Academic Press, New York).

STARBÄCK, K. (1895). *Bihang*, Discomyceten Studien, *till K. Sv. Vet.-Akad. Handl.*, **21** (iii), 5.

STOLK, AMELIA and SCOTT, DE B. (1967). Studies on the genus *Eupenicillium* Ludwig. I. Taxonomy and nomenclature of Penicillia in relation to their sclerotioid ascocarpic states. *Persoonia*, **4**, 391.

SUBRAMANIAN, C. V. (1962). A classification of the Hyphomycetes. *Curr. Sci.*, **31**, 409.

SUTTON, B. C. (1964*a*). Coelomycetes III. *C.M.I. Mycological Papers*, No. **97** (Commonwealth Mycological Institute, Kew).

SUTTON, B. C. (1964*b*). *Phoma* and related genera. *Trans. Br. mycol. Soc.*, **47**, 497.

TALBOT, P. H. B. (1954). Micromorphology of the lower Hymenomycetes. *Bothalia*, **6**, 249.

TALBOT, P. H. B. (1965). Studies of *Pellicularia* and associated genera of Hymenomycetes. *Persoonia*, **3**, 371.

TALBOT, P. H. B. (1968). Fossilized pre-Patouillardian taxonomy? *Taxon*, **17**, 620.

THAXTER, R. (1914). New or peculiar Zygomycetes. 3. *Blakeslea, Dissophora*, and *Haplosporangium* nova genera, *Bot. Gaz.*, **58**, 353.

TOWNSEND, BRENDA B. (1954). Morphology and development of fungal rhizomorphs. *Trans. Br. mycol. Soc.*, **37**, 222.

TOWNSEND, BRENDA B. and WILLETTS, H. J. (1954). The development of sclerotia of certain fungi. *Trans. Br. mycol. Soc.*, **37**, 213.

TUBAKI, K. (1958). Studies on the Japanese Hyphomycetes V. Leaf and stem group with a discussion of the classification of Hyphomycetes and their perfect stages. *J. Hattori Bot. Lab.*, **20**, 142.

VAN DER PLANK, J. E. (1963). *Plant Diseases: Epidemics and Control* (Academic Press, New York).

WAKEFIELD, ELSIE M. and BISBY, G. R. (1941). List of Hyphomycetes recorded for Britain. *Trans. Br. mycol. Soc.*, **25**, 49.

WARCUP, J. H. (1959). Studies on Basidiomycetes in soil. *Trans. Br. mycol. Soc.*, **42**, 45.

WARCUP, J. H. and TALBOT, P. H. B. (1962). Ecology and identity of mycelia isolated from soil. *Trans. Br. mycol. Soc.*, **45**, 495.

WARCUP, J. H. and TALBOT, P. H. B. (1966). Perfect states of some Rhizoctonias. *Trans. Br. mycol. Soc.*, **49**, 427.

WARCUP, J. H. and TALBOT, P. H. B. (1967). Perfect states of Rhizoctonias associated with orchids. *New Phytologist*, **66**, 631.

WEBSTER, J. (1959). Experiments with spores of aquatic Hyphomycetes I. Sedimentation and impaction on smooth surfaces. *Ann. Bot., N.S.*, **23**, 595.

WELLS, K. (1965). Ultrastructural features of developing and mature basidia and basidiospores of *Schizophyllum commune*. *Mycologia*, **57**, 236.

WHETZEL, H. H. (1945). A synopsis of the genera and species of the Sclerotiniaceae, a family of stromatic Inoperculate Discomycetes. *Mycologia*, **37**, 648.

WHITNEY, H. S. (1964). Physiological and cytological studies of basidiospore repetition in *Rhizoctonia solani* Kühn. *Can. J. Bot.*, **42**, 1397.

WILLIAMS, P. G., SCOTT, K. J. and KUHL, JOY L. (1966). Vegetative growth of *Puccinia graminis f.sp. tritici* in vitro. *Phytopathology*, **56**, 1418.

WILSENACH, R. and KESSEL, M. (1965). On the function and structure of the septal pore of *Polyporus rugulosus*. *J. gen. Microbiol.*, **40**, 397.

INDEX